an introduction to NUMERICAL METHODS WITH PASCAL

INTERNATIONAL COMPUTER SCIENCE SERIES

an introduction to NUMERICAL METHODS WITH PASCAL

L. V. ATKINSON

P. J. HARLEY

ADDISON-WESLEY PUBLISHING COMPANY

London · Reading, Massachusetts · Menlo Park, California · Amsterdam
Don Mills, Ontario · Manilla · Singapore · Sydney · Tokyo

Phototypesetting by Parkway Group, London and Abingdon
Printed in Finland by Söderström Osakeyhtiö, Member of Finnprint

British Library Cataloguing in Publication Data
Atkinson, Laurence
An introduction to numerical methods with Pascal.
(International computer science series; 6)
1. Numerical analysis—Data processing
2. PASCAL (Computer program language)
I. Title II. Harley, P. J. III. Series
519.4 QA297

ISBN 0-201-13788-7

Library of Congress Cataloging in Publication Data
Atkinson, Laurence.
An introduction to numerical methods with Pascal.
(The International computer science series)
Bibliography: p.
Includes index.
1. Numerical analysis—Data processing.
2. PASCAL (Computer program language)
I. Harley, P. J. 1948– . II. Title. III. Series.
QA297.A85 1983 519.4'028'5424 83–2520

ISBN 0-201-13788-7

ABCDEF 89876543

To
ANNE
and
JENNIFER

PREFACE

At the University of Sheffield, as at many other universities, numerical analysis is a major topic in undergraduate mathematics. Students are expected to carry out some hand calculations to acquire a basic understanding of numerical methods and, in order to solve more taxing problems and gain a deeper insight into the application of the methods, are also expected to implement them on a computer.

The main undergraduate programming language at Sheffield is Pascal and it has been our experience that students find difficulty in making the change from writing the programs expected in a general purpose Pascal programming course to writing programs for numerical methods. In this book, it is our intention to introduce some of the numerical methods forming the basis of those in current use, together with their implementation in Pascal. Rather than simply accompany each method by a Pascal routine, we have attempted to integrate programming aspects with the descriptions of the methods. All routines in the book are stored in a library file and are made available to the students. Copies of this file can be obtained from the authors.

We would like to acknowledge the assistance given to us during the preparation of this book. We are deeply indebted to Andrew McGettrick and Jan van Leeuwen, the series editors, for their many suggested improvements to early drafts. Thanks are also due to David 'Berto' Windle, of the University of Sheffield, for spotting some mistakes in part of the manuscript and to Jack Lambert of the University of Dundee for several helpful suggestions. An extremely convivial discussion with Jack laid the foundations for Chapter 8; it is obvious from the text that we found invaluable his excellent book on the numerical solution of ordinary differential equations. Any errors, omissions or lack of clarity naturally remain our responsibility and not that of those named above. Figure 6.2 is taken from Fig. 10.5 in 'Applied Numerical Analysis' by C. F. Gerald, 1978, Addison-Wesley, Reading, MA, and is reprinted with permission.

Finally, on the publishing side, we would like to extend thanks to Alan Whittle for copy editing 'above and beyond the call of duty' and to Sarah Mallen and Simon Plumtree for coping promptly with all queries and for maintaining a final submission date which was forever flexible.

Sheffield
February 1983

L. V. Atkinson
P. J. Harley

Contents

INTRODUCTION

The vast majority of mathematical models describing physical processes cannot be solved analytically. The only way to obtain any idea of the behaviour of a solution is to approximate the problem in such a way that numbers representing the solution can be produced. The process of obtaining the solution is then reduced to a series of arithmetic operations and such a process is called a *numerical method*.

The derivation and analysis of such methods lies within the discipline of *numerical analysis*. This book focusses on methods and most of the analysis is kept simple, concentrating on the derivation of the methods. Further analysis is presented when it is straightforward but results more difficult to obtain are quoted and references for further reading are given.

Traditionally numerical methods have been programmed in languages such as Fortran, Algol 60 (and 68) and PL/1. A question one might ask, therefore, is 'Why use Pascal?' One simple answer is that, at the higher educational level, Pascal has become the most widely used language for the teaching and illustration of good programming style and it is sensible that students of numerical methods should be on par with their computer science contemporaries. Further, Pascal is being used increasingly as an implementation language for practical applications on small computers and microcomputers and is thus becoming more widely available in all programming environments.

A more constructive argument arises from consideration of the reasons for Pascal's success. Pascal was designed to facilitate the teaching of good programming style; its principal merits are *transparency* (the intention of a well-written program is self-evident), *security* (many mistakes one might make can be detected by the computer and, in particular, by the compiler before a program is run) and *efficiency* (its design took implementation into account). The purpose of this book is not to present a collection of routines that might appear in a professional software library; rather, the intention is to introduce several numerical techniques and to encourage the reader to implement the methods on a computer so as to reinforce his understanding. It is important that the programs should be readable (transparent), it will be beneficial for the student to have as many mistakes detected as possible (security) and it is sensible that the programs run in a realistic time (efficiency).

One of Pascal's main advantages over most other programming languages lies in its provision of structuring facilities for both program control and data, and we shall make full use of these. Program control structures make a program more readable and, hence, its intended effect more apparent; data structures allow the programmer to group together related data items and this adds further to program readability. We shall occasionally present alternative approaches which may be necessary when using a programming language lacking some of Pascal's features but, for the most part, Pascal is exploited to the full. For readers less than fully familiar with Pascal, Chapter 1 presents a summary of the language and gives examples of some aspects particularly relevant to the implementation of numerical methods.

One fundamental feature of a numerical method is that it may not produce the 'right' answer. Two types of error are incurred: in general a numerical method produces an approximation to the right answer and this approximation may be affected by the rounding inherent in computer arithmetic. The nature and control of these errors must be understood and they are discussed in Chapter 2. Each following chapter deals with one particular class of numerical method.

Chapter 3 discusses the solution of non-linear equations. The chapter concentrates on the solution of a single equation in a single unknown but a final section introduces the problem of solving a *system* of non-linear equations. Chapter 4 discusses systems of linear equations: the solution of n linear equations in n unknowns. The progress of some of the methods of Chapter 4 depends upon the eigenvalues of the coefficient matrix and Chapter 5 presents techniques for determining both eigenvalues and eigenvectors.

Chapter 6 discusses problems of function approximation and Chapter 7 presents numerical differentiation and integration; the methods of Chapter 7 are often based on results given in Chapter 6. Chapter 8 introduces the numerical solution of ordinary differential equations and, as might be expected, has links with Chapter 7. The numerical solution of ordinary differential equations forms a vast field and space does not permit a comprehensive coverage in the present volume. Accordingly, this final chapter differs from the rest of the book in that it presents an overview of the methods rather than explaining each method in detail. Its aim is to give the flavour of the different types of method available.

This book assumes a working knowledge of Pascal and a familiarity with elementary undergraduate calculus and algebra. Having reached the end of the book, the reader new to numerical analysis should have a basic understanding of modern numerical techniques and should be able to program them in Pascal. The continued exposure to Pascal and its practical application should benefit all Pascal programmers.

Chapter 1 A SUMMARY OF PASCAL

This book assumes a familiarity with Pascal but, for completeness, we include here a summary of the language's features and an indication of the way in which we shall apply them to numerical methods. The language used conforms to the BSI/ISO standard (1982) and the style of Pascal adopted follows that of L. V. Atkinson (1980, 1981 and 1982).

A Pascal program has the following constituents

program heading
global label declarations
global constant declarations
global type declarations
global variable declarations
global procedure and function declarations
program body

All but the program heading and the program body are optional but any that are present must be in the relative position shown (e.g. constants must be declared before types, variables before procedures and functions). We now look briefly at each constituent.

1.1 Program body

The program body starts with the word **begin** and finishes with the word **end** and a period. Between this **begin** and **end** are the statements of the body, separated by semicolons.

Throughout this book, reserved words such as **begin** and **end** will appear in bold face. When you submit a program to a computer, you are advised to use capital letters for reserved words and lower case letters in most other contexts. This will make your programs more readable and, hence, their intended effect more transparent.

As a further aid to readability, you should include comments within your programs. A comment is a sequence of characters enclosed between { and } and is purely for the benefit of the human reader; the compiler treats a comment as though it were a space.

1.1.1 Output statements

If a file other than the standard file *output* is to be used it must be initialized for writing by a statement of the form *rewrite*(f). There are four output statements.

(i) *write* ($f, e_1, e_2, \ldots, e_n$)

outputs to the file f the values of the expressions $e_1, e_2, \ldots, e_n$. The file f may be of any component type not itself involving files and $e_1, e_2, \ldots, e_n$ must be of a type compatible with this component type. If f is a text file the values will be printed on one line and with an implementation-defined field width. If f is the standard file *output* then its name may be omitted from the write-statement.

(ii) *writeln* ($f, e_1, e_2, \ldots, e_n$)

differs from (i) in two respects: f must be a text file and an end-of-line marker will be written immediately after the value of e_n.

(iii) *put* (f)

writes to the file f the current value of the file buffer variable $f\uparrow$. The file f may be of any component type.

(iv) *page* (f)

places an end-of-page marker in the file f, which must be a text file. The effect of this statement varies depending upon the device subsequently used to print the file. For example, if the device is a VDU terminal it should clear the screen but, under some implementations, it may have no effect.

1.1.2 Input statements

If a file other than the standard file *input* is to be used it must be initialized for reading by a statement of the form *reset*(f). There are three input statements.

(i) *read* ($f, v_1, v_2, \ldots, v_n$)

reads n values from the file f and assigns them, in order, to the n variables $v_1, v_2, \ldots, v_n$. The file f may be of any component type not itself involving a file type and $v_1, v_2, \ldots, v_n$ must be of a type compatible with this component type. If f is a text file and integers or reals are being read then any spaces or end-of-line markers preceding a value are skipped. If the file is the standard file *input* then its name may be omitted from the read-statement.

(ii) *readln* ($f, v_1, v_2, \ldots, v_n$)

differs from (i) in two respects: f must be a text file and, after reading the value for v_n, all further characters are skipped up to, and including, the next end-of-line marker. It must be mentioned that some terminal implementations of Pascal are very bad in their handling of *readln*; they insist that a character be supplied after the end-of-line marker. If you suffer from such an implementation you will have to use *readln* with caution (and press your supplier for an improvement)!

(iii) *get* (*f*)

reads the next value from the file *f* and assigns it to the file buffer variable *f* ↑ . The file *f* may be of any component type not itself involving a file type.

1.1.3 Assignment-statement

The assignment-statement has the form

$v := e$

where v is a variable currently in scope, and
e is an expression whose type is compatible with the type of v.

v may be the name of a simple variable (e.g. x), a field of a record (e.g. *matrix.size*), a file buffer variable (e.g. *f* ↑), a pointer expression (e.g. *row* ↑), a subscripted variable (e.g. *a*[*i,j*]) or may involve a combination of subscripts, pointers and field selectors (e.g. *a*[*i*] ↑ .*coeff*[*j*]). When the assignment is encountered at run-time the address of the left-hand side variable is ascertained, the value of the right-hand side expression is computed and the value obtained is placed in the storage location with the specified address.

1.1.4 If-statement

In its full form this is

if b **then** s_1 **else** s_2

where b is a boolean expression (i.e. a test), and
s_1 and s_2 are any statements.

Its effect is to select one of two alternative statements. If b is *true* then statement s_1 is obeyed (and s_2 is not); if b is *false* then s_2 is obeyed (and s_1 is not). In either case, control then passes to the statement next in sequence after the if-statement.

It is commonly the case that s_2 is itself conditional and the statement following its **else** may also be conditional. This gives rise to the general form

if b_1 **then** s_1 **else**
  if b_2 **then** s_2 **else**
    if b_3 **then** s_3 **else**

whose effect is to make several tests in sequence (b_1, b_2, b_3, . . .) until one is *true*, whereupon the statement following the corresponding **then** is obeyed. If none of the tests is true, the statement after the final **else** is obeyed (if there is one).

There is a shortened form of the if-statement.

if *b* **then** *s*

where *b* is a boolean expression, and
s is any statement.

This merely determines whether a statement *s* will be obeyed or not.

1.1.5 Case-statement

This usually has the following form

case *e* **of**
  $v_1 : s_1$;
  $v_2 : s_2$;
  .
  .
  .
  $v_n : s_n$
end { *case* }

where *e* is an ordinal expression,
$v_1, v_2, \ldots, v_n$ are constants of the same ordinal type as *e*, and
$s_1, s_2, \ldots, s_n$ are any statements.

At run-time the expression *e* (called the selector) is evaluated and, assuming it evaluates to one of the quoted values v_i (called case labels), the appropriate statement s_i (called a case limb) is obeyed. A run-time error will result if the value of the selector is not present as one of the case labels. All labels within the case-statement must be distinct but several labels, separated by commas, may precede one limb.

1.1.6 Compound-statement

Any group of statements, separated by semicolons, can be bracketed between **begin** and **end**

begin
  s_1; s_2; . . .; s_n
end

to form one compound-statement. A compound-statement is used if several statements are to be obeyed when, otherwise, only one would be. Examples are as a limb of a case-statement or as a statement following **then**, **else** or **do**.

1.1.7 For-statement

There are two forms. An incremental for-statement has the form

for $v := e_1$ **to** e_2 **do** s

where v is a local ordinal variable,
e_1 and e_2 are ordinal expressions of a type compatible with v, and
s is any statement.

The statement is immediately exited if $e_1 > e_2$. Otherwise, the statement s (called the loop body) is obeyed a total of

$$ord(e_2) - ord(e_1) + 1$$

times with v (called the control variable) taking successive values

$$e_1, succ(e_1), succ(succ(e_1)), \ldots, e_2.$$

A decremental for-statement has the form

for $v := e_1$ **downto** e_2 **do** s

with v, e_1, e_2 and s as previously.

The statement is immediately exited if $e_1 < e_2$. Otherwise, the loop body is obeyed a total of

$$ord(e_1) - ord(e_2) + 1$$

times with the control variable taking successive values

$$e_1, pred(e_1), pred(pred(e_1)), \ldots, e_2$$

For both forms of for-statement, no attempt must be made to change the value of the control variable within the body of the loop and, upon natural exit from the loop, its value becomes undefined. Note that a for-loop control variable cannot have type *real*. In mathematical applications the type of the control variable is nearly always some integer subrange.

We shall use a for-statement when we know (or the program can easily determine) how many times we want a loop to be obeyed.

1.1.8 While-statement

This has the form

while b **do** s

where b is a boolean expression, and
s is any statement.

If b is *false* upon entry the while-statement is immediately exited. Otherwise, the loop body (s) is obeyed and the test b made again. This process is repeated until b eventually delivers *false* whereupon control leaves the while-statement.

We shall use a while-statement when we don't know in advance how many times we want a loop to be obeyed, and we may not want it obeyed at all.

1.1.9 Repeat-statement

This has the form

repeat
  s_1; s_2; . . .; s_n
until b

where b is a boolean expression, and
  $s_1, s_2, \ldots, s_n$ are any statements

and is similar to the while-statement but differs in three respects.

(i) The loop body may contain several statements without the need for **begin** and **end**.
(ii) The test is made *after* the loop body is obeyed rather than before and, consequently, the body of a repeat-loop is obeyed *at least once*.
(iii) Exit from the loop occurs when the test is *true*.

We shall use a repeat-loop when we don't know how many times a loop is to be obeyed, but it must be at least once.

1.1.10 With-statement

When records are being used several references are often made to fields of one record within a small group of statements. The with-statement offers a shorthand notation for these references. For example the statements

r.a := 17;   *r.b* := *true*;
r.c := '*X*';   *r.d* := '*Pascal*'

can be written

with *r* **do**
begin
  a := 17; *b* := *true*; *c* := '*X*'; *d* := '*Pascal*'
end

1.1.11 Goto-statement

The statement

goto *l*

where *l* is a statement label

transfers control directly to the statement labelled *l*. A statement label is an unsigned integer and the destination statement must be in the same block (procedure body, function body or program body) as the goto-statement or in a block textually enclosing the block containing the goto-statement. No attempt must be made to transfer control into a procedure body, function body, case-statement, if-statement, compound-statement, while-statement, repeat-statement or with-statement, but control can be transferred out of such constructs. If control is transferred directly out of a for-statement, the control variable retains the value it had immediately prior to the execution of the goto-statement.

The goto-statement is a particularly dangerous tool and its use is not to be encouraged. Program transparency and program reliability can suffer dramatically if goto-statements are introduced needlessly. Their use should be restricted to exiting from control structures in the event of an unexpected condition. In this book their sole purpose will be to handle exceptional conditions such as the breakdown of a numerical process (e.g. discovery of singularity of a matrix during attempted inversion). See Fig. 4.2 for an example of the use of a goto-statement.

1.2 Program heading

This has the form

program $p\ (f_1, f_2, \ldots, f_n)$

where p is the program name, and
$f_1, f_2, \ldots, f_n$ are the names of files used within the program.

Implementations vary but we shall assume that the two standard files *input* and *output* refer to the terminal and any other external files we introduce will be disk or tape files. The program heading is separated by a semicolon from the rest of the program.

1.3 Declarations

A program, function or procedure constitutes a *block* and entities may be declared local to a block. An entity is said to be *in scope* from its point of declaration to the end of the most locally enclosing block. If its identifier is redefined within an inner block, the original entity is not in scope within the inner block. Any declarations which occur must appear in the order listed at the start of this chapter and, because a declaration must always be separated from what follows by a semicolon, we can think of the semicolon as a terminator of the declaration.

1.3.1 Labels

Any statement label appearing in a program must be declared at the head of the most locally enclosing block. The labels are listed, separated by commas and preceded by a single occurrence of the reserved word **label**. For example,

label 999, 1;

1.3.2 Constants

User-defined constants may be of any scalar type or string type. A sequence of constant definitions is preceded by a single occurrence of the reserved word **const**. For example,

```
const
   pi = 3.14159;
   n = 25;
   circledegrees = 360;
   assumedzero = 1E-20;
   tempscale = 'C';
   title = 'Program to solve a quadratic equation';
```

We shall use constants to name mathematical constants (such as *pi*) and to name bounds of subranges. For example, if we denote the size of a square array by the constant *n*, we can define a subrange type 1 . . *n* for subscripts.

1.3.3 Types

Pascal provides four standard types

boolean — the truth values *true* and *false*
char — the set of characters available
integer — all whole numbers between two implementation-defined bounds −*maxint* and *maxint*
real — real numbers within an implementation-defined range and represented to an implementation-defined accuracy.

Computers represent information in binary form and binary digits (called *bits*) are grouped together to form what are called *words*. Typically, the number of bits in a word can vary from 16 to 60 depending upon the make of computer. The smallest addressable unit of storage is often a *byte* (eight bits) and, often, a word is two bytes. Inside the computer a boolean value requires one bit but usually occupies a byte or a word; a character requires a byte but may well occupy a word; an integer occupies one word (two bytes) or two words (four bytes); a real value usually occupies two words (four bytes) or four words (eight bytes).

On some computers multiple-precision arithmetic packages are available. These are collections of routines to carry out the standard arithmetic operations but using real numbers occupying more words than usual. They can be useful when precision of reals is very important but can slow a program down considerably.

In addition to the standard types, the user may define further types. To avoid confusing the names of types with other names in a program, we shall normally use plurals for user-defined types.

Symbolic types

Pascal's symbolic (or enumeration) types allow us to give names to the values we wish to process. For example,

```
days = (monday, tuesday, wednesday,
        thursday, friday, saturday, sunday);
```

Each symbolic value occupies one byte or one word depending upon the machine and implementation. The main application we shall make of these is to name states of interest within a computation. In practice, we shall frequently use symbolic types to control a multi-exit loop, that is, a loop with more than one reason for exit. For example, with the following iteration scheme

$$x_{n+1} = \frac{x_n + v/x_n}{2}$$

successive values x_{n+1} ($n = 0,1,2,\ldots$) approach the square root of v. We might terminate the process when the absolute difference between v and x_{n+1}^2 is less than some specified tolerance, say 10^{-7}.

```
const
  tolerance = 1E-7;

var
  x, v : real;
...
read (v, x);
while abs (sqr(x) - v) > tolerance do
  x := (x + v/x) / 2
```

If the tolerance specified is very small the process may take a long time to converge and, if the specified tolerance is beyond the limits of machine accuracy, the process may never converge at all. To guard against this, it is customary to count the number of iterations and terminate the process if this count reaches some specified bound. The loop now has *two* possible reasons for exit (tolerance achieved or maximum iterations reached) and so we have three states of interest:

(i) still computing new iterates,
(ii) tolerance achieved,
(iii) maximum iterations reached.

We can use a three-state scalar variable to control the loop

```
const
  tolerance = 1E-7;   maxits = 50;

var
  x, v : real;
  itcount : integer;
  state : (iterating, withintol, maxitsreached);
. . .

read (v, x);
itcount := 0;   state := iterating;
repeat
  if abs (sqr(x) - v) <= tolerance then state := withintol else
    if itcount = maxits then state := maxitsreached else
    begin
      x := (x + v/x) / 2;   itcount := itcount + 1
    end
until state <> iterating
```

and, upon exit from the loop, we can determine the current state with a case-statement.

```
case state of
  withintol     : . . .;
  maxitsreached : . . .
end { case }
```

We could use an if-statement

```
if state = withintol then . . . else . . .
```

but the case-statement is better; it gives more information. At this point in the program, the variable *state* can take one of only two values and the case-statement makes this fact explicit. Thus the case-statement makes the program more transparent.

Further, we may wish to guard against division by zero and terminate the iteration if x becomes too small. We would introduce a suitably small constant, *assumedzero*, and extend the number of states of interest to four.

```
state : (iterating, withintol, maxitsreached, rootnearzero);
```

We would then make a further test within the loop

```
if abs(x) <= assumedzero then state := rootnearzero else
```

and add a further limb to the case-statement.

```
case state of
  withintol    : . . .;
  maxitsreached: . . .;
  rootnearzero : . . .
end { case }
```

Subrange types
Scalar types other than *real* are called *ordinal* types. A subrange type is a subrange of the values defined by an ordinal type (which is then called the *host* type). For example,

```
letters = 'a' . . 'z';
degrees = 0 . . 360;
outcomes = withintol . . rootnearzero;
subs = 1 . . n;
```

In the square root fragment above, the counting of loop iterations could be performed by a subrange variable.

```
itcount : 0 . . maxits;
```

Any attempt to assign to a subrange variable any value outside the permitted range constitutes an error so, by utilising subrange types, we are helping the computer to detect our mistakes. In particular, it is often possible for the *compiler* to detect subrange errors and this compile-time security is one of Pascal's greatest assets.

Arrays
The array is Pascal's equivalent of the mathematician's vector.

```
const
  n = 20;

type
  subs = 1 . . n;
  vectors = array [subs] of real;
```

Associated with any array is a component type (*real* in the above example) and an index type (*subs* in the above example). To identify one particular element of an array, the array name is followed by a subscript inside square brackets. For example,

```
var
   u, v : vectors;
   i : subs;
   dot : real;
...
dot := 0;
for i := 1 to n do
   dot := dot + u[i] * v[i]
```

The index type of an array must be an ordinal type but the component type may be any type and, in particular, it may be an array type. This gives a multidimensional array type. For example,

```
const
   n = 25;

type
   subs = 1 . . n;
   vectors = array [subs] of real;
   nbyn = array [subs] of vectors;

var
   a, b : nbyn;
```

The variables a and b are $n \times n$ matrices and so the expression $a[i]$ denotes a vector and an element of this vector can be identified by appending a further subscript: $a[i][j]$. As a notational convenience, the adjacent square brackets may be replaced by a comma

```
a[i,j] is equivalent to a[i][j]
```

Correspondingly, a shorthand notation is applicable to array declarations in that the data type

```
nbyn = array [subs, subs] of real
```

is equivalent to the type

```
nbyn = array [subs] of array [subs] of real
```

The structures defined by these two types are effectively the same as those defined by the original data type *nbyn* but the original form is to be preferred because it enables us to do more with *slices* of an array. The reason for this need not concern us here but, in Section 1.6, we shall look at slicing in some detail. Throughout this book we shall often assume the existence of the following definitions.

```
const
  n = . . .;   { a positive integer }

type
  subs = 1 . . n;
  vectors = array [subs] of real;
  nbyn = array [subs] of vectors;
```

Records

An array is a collection of variables all of the same type, each identified by a subscript (or set of subscripts) evaluated at run-time. A record is a collection of variables, not necessarily of the same type, each identified by a *field selector*, and this selector is known to the compiler. For example, if we wish to process complex numbers, we can define a record type to represent a complex number.

```
complexnos = record
               re : real;
               im : real
             end { complex nos };
```

As with ordinary variable declarations, a shorthand notation is available for record field selectors of identical type. The record type above can be defined as follows

```
complexnos = record
               re, im : real
             end { complex nos };
```

Given a variable of this type, a particular field is selected by following the variable by the name of the field identifier and separating the two by a dot. As an illustration, the following program fragment forms the sum and product of two complex numbers.

```
var
  c1, c2, s, p : complexnos;
. . .
s.re := c1.re + c2.re;   s.im := c1.im + c2.im;
p.re := c1.re * c2.re − c1.im * c2.im;
p.im := c1.re * c2.im + c1.im * c2.re
```

Data structure definitions should not pre-empt the program body. An algorithm should be defined independently of the data structures and then data representations chosen to best facilitate the algorithm.

Files

The two standard files *input* and *output* are *text* files. These are files of characters, broken into lines by the inclusion of end-of-line markers. The data type *text* is predefined and any non-standard text file quoted in the program heading must be declared in the main block of the program as a variable of this type.

The next character waiting to be read from a text file *f* is available as the value of the file buffer variable *f*↑, often called the file window. If the character in the file window is the end-of-line marker, the expression

eoln (*f*)

delivers *true*; otherwise this expression delivers *false*. If the marker at the end of the very last line of the file *f* has been passed, the expression

eof (*f*)

delivers *true*; otherwise this expression delivers *false*.

Files of component type other than *char* may be constructed and this is particularly useful for numerical analysis programs. Data supplied to a program is usually in the form of numbers rather than arbitrary characters and these numbers are often regarded as being grouped in some way. For example, if one program produces a 10 × 10 matrix which a second program will subsequently process, we can use the following file type.

```
const
  n = 10;

type
  subs = 1 . . n;
  vectors = array [subs] of real;
  nbyn = array [subs] of vectors;
  nbynfiles = file of nbyn;
```

The first program then writes to a file of this type

```
program prog1 (input, output, matfile);
  . . .
  var
    matfile : nbynfiles;
    mat : nbyn;
  . . .
  rewrite (matfile);   write (matfile, mat)
```

and the second program reads from it.

```
program prog2 (input, matfile, output);
. . .
var
   matfile : nbynfiles;
   mat : nbyn;
. . .
reset (matfile);   read (matfile, mat)
```

An implementation may limit the number of words involved in one file transfer, and a typical limit might be 1024 (2^{10}). In this case and on a machine where real values occupy two words, the above example would not work for $n>16$. The program would have to use a **file of** *vectors* and transfer the matrix one row at a time.

Sets

In Pascal, a set is a collection of values of some ordinal type and with ordinal numbers within some implementation-defined range (typically 0 to 127). For example,

```
const
   n = 25;

type
   letters = 'a' . . 'z';
   subs = 1 . . n;

var
   vowels : set of letters;
   subscripts : set of subs;
   k : integer;
. . .
   vowels := ['a', 'e', 'i', 'o', 'u'];
   subscripts := [1 . . n];
```

Set inclusion is tested with the operator **in**. For example,

```
if k in subscripts then . . .
```

We shall use sets in only one program (Fig. 6.9).

Pointers

A pointer is the Pascal name for the *address* of an object within the computer. A pointer occupies one word. Pointers are particularly useful when large data items are to be reordered in some way because, rather than move the data items themselves, we can set up pointers to the items and move these. For example, if we wish to interchange rows of a matrix M and the corresponding elements of a vector v, we can utilise the following data types.

```
const
  n = . . .;

type
  subs = 1 . . n;
  vectors = array [subs] of real;
  rows =
    record
      vec : vectors;
      v : real
    end { rows };
  rowptrs = ↑ rows;
  ptrvecs = array [subs] of rowptrs;
```

The interchange of rows 1 and 2 then has the following form

```
var
  m : ptrvecs;
  p : rowptrs;
. . .
p := m[1];  m[1] := m[2];  m[2] := p
```

and performs three assignments, each involving the copying of a single word. Using the first data types defined for matrices and vectors the operation would be considerably more expensive.

```
var
  m : nbyn;
  temprow, v : vectors;
  tempreal : real;
. . .
temprow := m[1];  m[1] := m[2];  m[2] := temprow;
tempreal := v[1];  v[1] := v[2];  v[2] := tempreal
```

In this case, each of the first three assignments involves n real values.

1.4 Variables

Variables should always be declared as locally as possible. If a variable is used only in one routine then it should be declared local to that routine. Do not confuse the use of the same identifier in two different routines with the use of the same variable in two different routines. A routine to form the dot product of two vectors will use a for-loop control variable, possibly called *i*,

```
dot := 0;
for i := 1 to n do
  dot := dot + u[i] * v[i]
```

and a routine to square all the elements of a vector might well use a control variable of the same name

```
for i := 1 to n do
  u[i] := sqr (u[i])
```

but these are two completely different variables. This is illustrated by the fact that we can change the name of one of them and the effect of the program will be unchanged. So each routine should have its own local declaration of the variable *i*. In fact, because these are for-loop control variables, the Pascal standard dictates that they *must* be declared locally (see Section 1.1.7).

1.5 Procedures and functions

A routine (a procedure or a function) is a section of program which has been given a name, possibly has some parameters associated with it and, if it is a function, delivers a value of some specified type. Pascal's standard functions are listed in Appendix 1. Throughout this book we shall be concerned with routines rather than with whole programs. Almost every numerical method described will be accompanied by a routine (usually a procedure) illustrating its implementation. For some methods more than one routine will be supplied, each illustrating a different approach. Most routines will be self-contained, that is, they will take as (value) parameters any information they require and will return via (variable) parameters any results they compute. One exception to this will be in the case of some constants that might be expected to be global to the whole program. An example of this is the constant *n* used in Section 1.3.3 (p.13) to denote the size of a vector or matrix. Another exception may occur when one routine is declared within another; the inner one may refer to a parameter of the outer one without taking it as a parameter.

One area to which we pay particular attention is that of side-effects. Any non-local variable whose value may be changed within a procedure body is made a variable parameter of the procedure. A function should cause no side-effect; that is, any changes caused by computation within the function body should be local to the function. One consequence of this is that, throughout this book, no function takes a variable parameter.

The square root example of Section 1.3.3 (p.10) could be cast as the procedure of Fig. 1.1. This assumes the existence of the subrange type

itcounts = 0 . . *maxits*;

and illustrates a typical parameter list for an iterative process. The routine is supplied with a value (*v*) whose root is sought, an initial approximation to the root (*x*), a tolerance (*tol*) and a bound on the number of iterations (*itbound*) and it indicates the outcome of the process via a boolean variable parameter (*success*). If *success* is *true*, the procedure returns the latest estimate of the root (*root*) and the number of iterations (*noofits*). Note that the parameter list may well be independent of the method. The procedure of Fig. 1.1 could apply a totally different algorithm but utilise the same parameters.

```
procedure FindSqRoot (v, x, tol : real;   itbound : itcounts;
                          var success : boolean;
                          var root : real;   var noofits : itcounts);
  var
    itcount : itcounts;
    state : (trying, doneit, toomanyits);

begin
  itcount := 0;   state := trying;
  repeat
    if abs (sqr(x) - v) <= tol then state := doneit else
      if itcount = itbound then state := toomanyits else
      begin
        x := (x + v/x) / 2;   itcount := itcount + 1
      end
  until state <> trying;
  success := state = doneit;
  if success then
  begin
    root := x;   noofits := itcount
  end
end {Find sq root};
```

Figure 1.1

Some processes require more than one starting value (perhaps two end points of a range or a vector of starting values), some require further information to enable them to carry out the method (perhaps a function or an acceleration factor) and some may have more than two possible outcomes (perhaps converged, iterations exceeded or point of inflexion encountered in a function). The procedures for such processes have suitably modified parameter lists. The procedure of Fig. 1.1 could be modified to take into account the fact that the iteration may break down when x is close to zero. There are then, not two, but three possible outcomes and so a variable boolean parameter does not suffice. Instead, we would define a named symbolic type to represent all the states of interest within the iteration loop and a named subrange type to identify the possible outcomes.

```
states = (iterating, withintol, maxitsreached, rootnearzero);
outcomes = withintol . . rootnearzero;
```

The procedure then has the form shown in Fig. 1.2.

There are, of course, other ways of handling error situations; for example, the procedure could print a message (and perhaps some numerical values) and then exit. However, different circumstances might call for different information to be displayed and so we shall maintain the convention adopted above; a variable parameter indicates any forseeable breakdown of a process and it is the job of the calling segment to act upon this.

```
procedure FindSqRoot (v, x, tol : real;   itbound : itcounts;
                      var outcome : outcomes;
                      var root : real;   var noofits : itcounts);
  var
    itcount : itcounts;
    state : states;
begin
  itcount := 0;   state := iterating;
  repeat
    if abs (sqr(x) - v) <= tol then state := withintol else
      if itcount = itbound then state := maxitsreached else
        if abs(x) <= assumedzero then state := rootnearzero else
        begin
          x := (x + v/x) / 2;   itcount := itcount + 1
        end
  until state <> iterating;

  outcome := state;
  if state = withintol then
  begin
    root := x;   noofits := itcount
  end
end { Find sq root };
```

Figure 1.2

```
program RootFinder (infile, output);

  { Finds the square root of all the numbers in a supplied file. The file also
    contains a starting value, tolerance and iteration bound for each
    number. }

  const
    maxits = 50;

  type
    itcounts = 0 . . maxits;

  var
    infile : text;
    number, startval, sqroot, tolerance : real;
    its, bound : itcounts;
    solved : boolean;

  procedure FindSqRoot
       (v, x, tol : real;   itbound : itcounts;
        var success : boolean;
        var root : real;   var noofits : itcounts);
    . . . { as in Fig. 1.1 }

begin { program body }
  writeln ('  number   starting   tolerance   iteration   root   iterations');
  writeln ('             value                  bound');
  writeln;
  writeln ('   number starting tolerance iteration root iterations');
  writeln ('             value              bound');
  writeln;
  reset (infile);
  repeat
    readln (infile, number, startval, tolerance, bound);
    write (number:10:6, startval:10:6, '    ', tolerance:8, bound:7);
    FindSqRoot (number, startval, tolerance, bound, solved, sqroot, its);
    if solved then writeln (sqroot:13:6, its:8) else
      writeln ('          -          -')
  until eof (infile);
  writeln;   writeln ('All supplied numbers have been processed')
end.
```

Figure 1.3

When testing programs, especially iteration programs, it is customary to print intermediate results at frequent stages throughout the computation. Although the reader is encouraged to do this, it will not be the practice in the routines presented in this book. As a general rule, information computed by a routine will be returned via one or more parameters and the printing of results will be the task of the calling segment. It is probable that the output generated by a program will be read by several people other than the original author. It is important therefore that the output results are clearly displayed and any tabulated information is presented beneath suitable column headings. A program which generates a few numbers but no explanatory text will be of little use to the author a few weeks after it has been written and will be of even less use to someone who has never seen the program before.

A program using the procedure of Fig. 1.1 might have the form shown in Fig. 1.3. When supplied with the following data:

9	1	1E−5	25
0.12345	0.5	1E−6	10
512	10	1E−10	5

this program produces the output shown in Fig. 1.4.

This program finds the square root of a sequence of values supplied in an external file

number	*starting value*	*tolerance*	*iteration bound*	*root*	*iterations*
9.000000	1.000000	1.0E−05	25	3.000000	5
0.123450	0.500000	1.0E−06	40	0.351355	3
512.000000	10.000000	1.0E−10	5	–	–

All supplied numbers have been processed

Figure 1.4

1.5.1 Parameter transfer

Programming aesthetics dictate that a parameter should be transferred as a variable only if an updated value is to be returned. However, when dealing with structures such as records and arrays, program efficiency can sometimes be significantly improved by contravening this principle.

The transfer of a value parameter differs from the transfer of a variable parameter. When a value parameter is transferred, a local copy is constructed and, within the procedure body, all references to the parameter become references to the local copy. This affects both time and space: extra storage is required and the copying process takes time. At the machine-code level, the copying of an array or record is achieved by some 'block move' operation rather than element by element but, for a large structure, the time involved may be significant, particularly if this copying is carried out many times. On the other hand, access to a local variable is marginally more efficient than access to a non-local variable. Thus, once the copying of a value parameter is completed, each reference to it should be marginally faster than a corresponding reference to a variable parameter.

If store is at a premium, a parameter may have to be transferred as a variable. If not, and if the parameter is not updated within the routine, there is freedom to choose the transfer mechanism. The factors governing the time are the size of the structure being transferred and the number of run-time references to its elements and so the important statistic is the ratio of the two. For any particular implementation, estimates can be obtained of the average time saving achieved by referencing a local variable and of the average copying overhead for a value parameter and the ratio of the two can be determined. However, the copying overhead is likely to be so much greater than the saving achieved by local reference that, in a practical situation, variable transfer is almost certain to be faster. Perhaps a more useful consideration is the ratio of the copying overhead to the total time of the processing carried out by a routine. If it is small, then value transfer is worth retaining; if it is large, then the saving achieved by variable transfer becomes significant.

Throughout this book we shall ignore this aspect of efficiency and, in the interests of good programming style, we shall use a variable parameter only when its value is to be updated by the routine.

1.5.2 Formal routines

Occasionally we wish to transfer one routine as a parameter to another. For example a procedure to find a zero of a function would take the function as a parameter. Its heading might be as follows:

procedure *FindZero* (**function** *f*(*x* : *real*) : *real*; **var** *root* : *real*);

or, perhaps, if it employs an iterative approach,

```
procedure FindZero (function f(x : real) : real;   itbound : itcounts;
                     var success : boolean;
                     var root : real;   var noofits : itcounts);
```

In both cases the parameter *f* is a formal function; when the procedure is called, the name of an actual function, in scope in the calling block, must be supplied. The actual function and the formal function must have the same number and types of parameters and must deliver a result of the same type.

1.5.3 Conformant arrays

When an array is passed as a parameter to a routine, the type of the actual array supplied must be identically that specified for the formal array. This constraint, together with the availability of subrange types, contributes to both security and efficiency; most, if not all, subscript checking can usually be performed by the compiler and efficient array access code can be generated because the array bounds are known.

However, the constraint does mean that a procedure taking an array parameter cannot be applied, at two different points in a program, to two arrays which have been declared with different index types. In an attempt to overcome this, we can specify a maximum size for the arrays and supply the procedure with extra parameters to indicate which portion of the array is to be processed during any particular activation. This is illustrated by the function of Fig. 1.5.

```
function dotprod (u, v : vectors;   m : subs) : real;
  var
    i : subs;
    dot : real;
begin
  dot := 0;
  for i := 1 to m do
    dot := dot + u[i] * v[i];
  dotprod := dot
end { dot prod };
```

Figure 1.5

Although all vectors supplied as actual parameters must have the same type (*vectors*) and hence be of the same length, we can process different vectors as though they were of different lengths by supplying different values for *m*. Assuming the following declarations

```
const
  n = 100;

type
  subs = 1 . . n;
  vectors = array [subs] of real;

var
  a, b, c, d : vectors;
```

the two function calls

dotprod $(a, b, 10)$

and

dotprod $(c, d, 74)$

are legal. In the first case the function computes

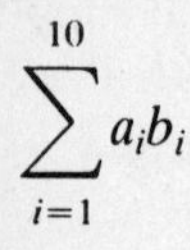

$$\sum_{i=1}^{10} a_i b_i$$

and, in the second

$$\sum_{i=1}^{74} c_i d_i$$

This approach is perfectly adequate for the individual programmer for most applications. However, it is inconvenient when a routine is to be made available to many users, each with a different application in mind, and so Pascal allows *conformant* array parameters. When specifying the type of a formal array parameter we need not use a named array type; instead we specify the component type, the number of dimensions and the host type of each index type. We must also supply identifiers which, inside the routine, will be used to denote the bounds of the dimensions of the actual parameter. This is illustrated by the function of Fig. 1.6.

```
function dotprod (u, v : array [lb. .ub:integer] of real) : real;
  var
    i : integer;
    dot : real;
begin
  dot := 0;
  for i := lb to ub do
    dot := dot + u[i] * v[i];
  dotprod := dot
end { dot prod };
```

Figure 1.6

The two actual parameters must be of the same one-dimensional array type but no constraint is placed upon the index type other than it must be a subrange of *integer*. Inside the procedure, *lb* identifies the lower bound of this subrange and *ub* identifies the upper bound. Given the following declarations

```
const
   n = 30;  k1 = -40;  k2 = 23;

type
   nsubs = 1 . . n;   ksubs = k1 . . k2;
   nvecs = array [nsubs] of real;
   kvecs = array [ksubs] of real;

var
   na, nb : nvecs;
   ka, kb : kvecs;
```

the two function calls

dotprod (*na*, *nb*)

and

dotprod (*ka*, *kb*)

are legal. For the first call, *lb* is taken to be 1 and *ub* is taken to be 30; for the second call, *lb* is −40 and *ub* is 23.

When a routine takes a conformant array parameter there can be a loss of both security and efficiency. Security can suffer because, inside the routine, less is known about the bounds of the array and efficiency suffers because the bounds of the array must be passed across dynamically when the routine is called. It should also be noted that a compiler need not accept conformant arrays. The standard admits two levels of Pascal and the conformant array facility forms part of the *level 1* standard. A compiler implementing only the *level 0* standard will not accept conformant arrays.

This book is aimed at the student of numerical methods rather than the implementor of a software library and so the programs are written to conform to the level 0 standard. Consequently, conformant arrays will not be used.

1.5.4 Recursion

A routine may call itself and is then said to be *recursive*. Recursion is a powerful device and, in some application areas, is an extremely useful tool. In general, numerical methods have little need of recursion but we shall find it very useful for one particular method of Chapter 7. We illustrate the concept first with an example chosen for its simplicity: the factorial function. The usual definition is:

$$n! = n(n-1)(n-2) \cdot \cdot \cdot (2)(1) = \prod_{i=1}^{n} i$$

and this gives rise to the following Pascal function which assumes that the data type *posints* is the integer subrange 1 to *maxint*.

```
function factorial (n : posints) : posints;
  var
    i, f : posints;
begin
  f := 1;
  for i := n downto 1 do
    f := f * i;
  factorial := f
end { factorial };
```

We should be equally happy with the following definition

$$n! = \left. \begin{array}{c} 1 \\ n \times (n-1)! \end{array} \right\} \begin{array}{l} n=1 \\ n>1 \end{array}$$

and this is recursive because the function is defined 'in terms of itself'. More correctly, it is defined in terms of a *subcase* of itself: *factorial n* is defined in terms of *factorial* $n-1$. Because Pascal allows a function to call itself we can model a Pascal function directly on this definition. A value (in our case $n-1$), computed at one recursive level, is passed as a parameter to the next.

```
function factorial (n : posints) : posints;
begin
  if n=1 then factorial := 1 else
    factorial := n * factorial (n-1)
end { factorial };
```

To confirm that this works, consider the call *factorial* (3).

This is computed as 3 × *factorial*(2)
where *factorial*(2) is 2 × *factorial*(1)
and *factorial*(1) is defined to be 1.

The value computed is therefore 3 × (2 × (1)) which is 6.

For the above example the non-recursive function is straightforward. This suggests that recursion is not appropriate and, indeed, conventional compilers produce less efficient code for the recursive form. To illustrate a more realistic application of recursion we consider an example for which the non-recursive version is significantly more cumbersome than its recursive counterpart. We shall write a procedure to print an integer in some base other than 10.

The decimal system uses base 10 with digits 0, 1, 2, 3, 4, 5, 6, 7, 8 and 9. When we write the decimal integer

30851

this is a shorthand notation for

$$3\times10^4 + 0\times10^3 + 8\times10^2 + 5\times10 + 1$$

The octal system, for example, uses base 8 and digits 0, 1, 2, 3, 4, 5, 6, 7 and the octal equivalent of the decimal value above is

74203

We can show this by transforming 74203_8 (where the suffix 8 indicates the base) into decimal.

$$\begin{aligned} 74203_8 &= 7\times8^4 + 4\times8^3 + 2\times8^2 + 0\times8 + 3 \\ &= 28672_{10} + 2048_{10} + 128_{10} + 0 + 3 \\ &= 30851_{10} \end{aligned}$$

If we repeatedly divide a positive integer by 8 until the dividend is reduced to 0, the remainders produced are the digits of the octal representation, but in reverse order. The decimal integer 30851 is treated in this way in Fig. 1.7.

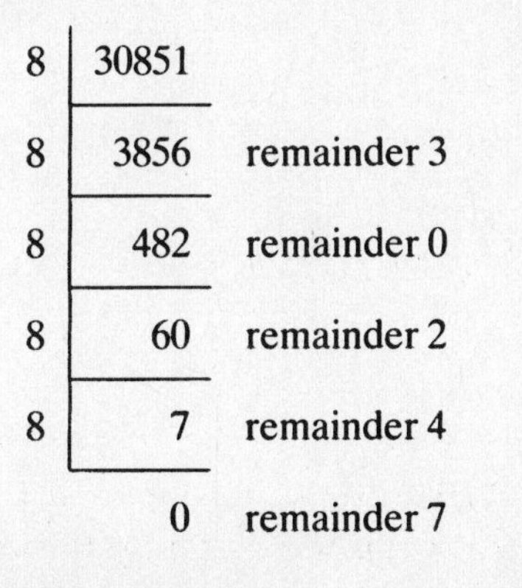

8 30851
8 3856 remainder 3
8 482 remainder 0
8 60 remainder 2
8 7 remainder 4
0 remainder 7

Figure 1.7

A procedure to print a positive integer in octal could utilise a local array to store successive remainders and, when the division process terminates, output them in reverse order. A much simpler procedure results if we use recursion: we define the octal printing process 'in terms of itself'. We describe the action

print in octal (*n*)

as the two steps

print in octal (*n* **div** 8);
write (*n* **mod** 8 :1)

unless n=0 when the process terminates. The procedure *PosPrint* of Fig. 1.8 is modelled directly on this and can be used for any base in the range 2 to 10. The technique is applicable to bases greater than 10 but then special symbols must be printed to represent 'digits' greater than 9. The call

PosPrint (479, 8)

prints the decimal value 479 in octal and the call

PosPrint (479, 2)

prints the same value in binary.

```
const
  minbase = 2;   maxbase = 10;

type
  bases = minbase . . maxbase;
  nonnegints = 0 . . maxint;

procedure PosPrint (n : nonnegints;   b : bases);
begin
  if n > 0 then
  begin
    PosPrint (n div b,b);   write (n mod b : 1)
  end
end { Pos print };
```

Figure 1.8

To accommodate negative and zero n we can encase this recursive process within a non-recursive procedure which recognises the additional cases. This is done in Fig. 1.9. The reader is encouraged to write a procedure equivalent to *Print* of Fig. 1.9 but without involving recursion. This exercise should prove the power of recursion.

```
procedure Print (m : integer;   b : bases);

   procedure PosPrint (n : nonnegints;   b : bases);
   begin
      if n > 0 then
      begin
         PosPrint (n div b, b);   write (n mod b : 1)
      end
   end { Pos print };

begin { Print }
   if m > 0 then PosPrint (m, b) else
      if m < 0 then
      begin
         write (' -');   PosPrint (-m, b)
      end else
         write ('0')
end { Print };
```

Figure 1.9

1.6 Slicing

Many numerical methods involve the processing of matrices and we shall meet some of these in Chapters 4 and 5. A matrix is a two-dimensional array of values but we often process only one or two rows or columns at a time. In such a circumstance, program efficiency can be improved by *slicing* the matrix to produce the desired rows or columns. There are essentially three ways to slice a matrix in Pascal and these will be discussed in turn, but first we consider the usual representation of a matrix and apply some simple matrix manipulations.

We shall consider only square matrices, but all three slicing techniques are applicable to rectangular matrices. We shall specify the matrix size as a user-defined constant

```
const
   n = . . .;
```

and adopt the definition of *nbyn* introduced in Section 1.3.3 (p.13).

```
type
  subs = 1 . . n;
  vectors = array [subs] of real;
  nbyn = array [subs] of vectors;
```

We now consider procedures to add and to multiply two matrices. The sum of two matrices A and B is a matrix C with the property that

$$c_{ij} = a_{ij} + b_{ij} \qquad (i, j = 1, 2, \ldots, n)$$

The product of two matrices A and B is a matrix C with the property that

$$c_{ij} = \sum_{k=1}^{n} a_{ik} b_{kj} \qquad (i, j = 1, 2, \ldots, n)$$

The procedures are in Figs. 1.10 and 1.11.

```
procedure AddMats (a, b : nbyn;   var c : nbyn);
  var
    i, j : subs;
begin
  for i := 1 to n do
    for j := 1 to n do
      c[i,j] := a[i,j] + b[i, j]
end { Add mats };
```

Figure 1.10

In these procedures, each reference to a matrix element takes the form of a doubly subscripted expression. Access efficiency is improved if we can use one subscript rather than two and this is the purpose of slicing. Because we have declared the data type *nbyn* in terms of *vectors*, we can extract entities of type *vectors* from any *nbyn* object. Thus we can slice a matrix — but only one way. If we think of the first subscript indicating a row and the second a column we can slice along rows but not down columns; that is, we can pick out a row but not a column. For any algorithm where we wanted to process columns rather than rows, we would use the first subscript to represent a column and the second a row — this is equivalent to storing the transpose of the matrix. We now examine the various techniques for slicing along rows.

```
procedure MultiplyMats (a, b : nbyn;   var c : nbyn);
  var
    i, j, k : subs;
    sigma : real;
begin
  for i := 1 to n do
    for j := 1 to n do
    begin
      sigma := 0;
      for k := 1 to n do
        sigma := sigma + a[i,k] * b[k,j];
      c[i,j] := sigma
    end
end { Multiply mats };
```

Figure 1.11

1.6.1 Local copying

If we have a local variable of type *vectors* we can assign to it any row of a matrix and thereby create a local copy of that row. Elements of the row can then be accessed by single subscripting. Correspondingly, to generate a row of a matrix, we can first generate a local copy and then use this copy to overwrite the appropriate row of the matrix. The procedure of Fig. 1.12 is a modification of the procedure of Fig. 1.10 to incorporate slicing. All three matrices are processed row-wise and so all three can be sliced.

```
procedure AddMats (a, b : nbyn;   var c : nbyn);
  var
    i, j : subs;
    ai, bi, ci : vectors;
begin
  for i := 1 to n do
  begin
    ai := a[i];   bi := b[i];
    for j := 1 to n do
      ci[j] := ai[j] + bi[j];
    c[i] := ci
  end
end { Add mats };
```

Figure 1.12

If we apply this technique to the procedure of Fig. 1.11, we can apply slicing only to *a* and *c* because *b* is processed column-wise (its first subscript varies more rapidly than the second). The resultant procedure is in Fig. 1.13.

```
procedure MultiplyMats (a, b : nbyn;   var c : nbyn);
  var
    i, j, k : subs;
    sigma : real;
    ai, ci : vectors;
begin
  for i := 1 to n do
  begin
    ai := a[i];
    for j := 1 to n do
    begin
      sigma := 0;
      for k := 1 to n do
        sigma := sigma + ai[k] * b[k,j];
      ci[j] := sigma
    end;
    c[i] := ci
  end
end { Multiply mats };
```

Figure 1.13

If *b* stored the transpose of a matrix *B* then *b* would be processed row-wise and so could be sliced. The procedure of Fig. 1.14 transposes *B* and then applies slicing to all three arrays. Notice that, because *b* is a value parameter of *MultiplyMats*, the actual parameter supplied will not be corrupted by the local transposing of *b*. The time taken to transpose the matrix will almost certainly be greater than any time saved by slicing *b* and so this procedure is presented merely to illustrate the slicing; unless the transpose of *B* is already available, the procedure of Fig. 1.13 is to be preferred.

The construction of a local copy of a row enables elements of that row to be accessed by a single subscript rather than two. This reduces the access time but, weighing against this, account must be taken of the time (and space) needed to construct the local copy. The time taken to construct the copy is proportional to the number of words copied and the total time saved by single subscripting is proportional to the number of references made to the elements of the row copied. The important statistic is therefore the ratio of the two: the average number of accesses per word. We shall call this the *word access ratio*.

In the procedure of Fig. 1.12 each element of each row copied is accessed

```
procedure MultiplyMats (a, b : nbyn;   var c : nbyn);

  var
    i, j, k : subs;
    sigma : real;
    ai, bj, ci : vectors;

  procedure Transpose (var m : nbyn);
    var
      i, j : subs;
      mij : real;
  begin
    for i := 2 to n do
      for j := 1 to i-1 do
      begin { Swap m[i,j] and m[j,i] }
        mij := m[i,j];   m[i,j] := m[j,i];   m[j,i] := mij
      end
  end { Transpose };

begin { Multiply mats }
  Transpose (b);
  for i := 1 to n do
  begin
    ai := a[i];
    for j := 1 to n do
    begin
      sigma := 0;   bj := b[j];
      for k := 1 to n do
        sigma := sigma + ai[k] * bj[k];
      ci[j] := sigma
    end;
    c[i] := ci
  end
end { Multiply mats };
```

Figure 1.14

once and so, if a real value occupies η words, the word access ratio for each of *ai*, *bi* and *ci* is $1/\eta$. The procedure of Fig. 1.13 makes n^2 accesses to elements of each *ai* and n accesses to elements of each *ci* and, because each row contains n elements, the word access ratio for *ai* is n/η and for *ci* is $1/\eta$. In Fig. 1.14, the word access ratio for a_i is n/η and for both b_j and c_i is $1/\eta$.

The obvious question now is 'How big does the word access ratio have to be to make slicing worthwhile?' To compare the saving achieved by single subscripting with the overhead incurred by copying a row we adopt the following definitions.

n = number of elements in each row
η = number of words occupied by each element
T_1 = time taken by a singly subscripted reference ($ai[j]$)
T_2 = time taken by a doubly subscripted reference ($a[i,j]$)
$S = T_2 - T_1$ = saving per reference ($S > 0$)
C = average time taken to copy one word (time to copy a row divided by $n\eta$)
r = word access ratio (number of accesses to elements of the row divided by n)

If a row contains n elements, each of η words, we can express the total saving as $n\eta rS$ and the total copying overhead as $n\eta C$. We are interested in the situation where the saving exceeds the overhead:

$$n\eta rS > n\eta C$$

This gives

$$rS > C$$

and so

$$r > C/S$$

The values of C and S vary from one implementation to another but, if your Pascal system provides a mechanism for timing sections of program (and most do) you can run some simple programs to obtain estimates. For large n, typical values for C/S lie in the range 0.1 to 0.5. It can be seen, therefore, that the slicing in Fig. 1.12 will have a significant effect only if both C/S is small (~0.1) and η is small (say, 2). On the other hand, the slicing of a in Figs. 1.13 and 1.14 should produce a marked improvement on any machine.

1.6.2 Record field selection

If we use records to represent a matrix we can use with-statements to achieve slicing.

```
const
  n = . . .;

type
  subs = 1 . . n;
  rows =
    record
      v : array [subs] of real
    end { rows };
  rowvecs = array [subs] of rows;
```

```
var
  d : rowvecs;
  i, j : subs;
```

In the context of the above declarations, we can slice *d* by using

```
with d[i] do
```

and so, for example, we can double every value in row *i* as follows:

```
with d[i] do
  for j := 1 to n do
    v[j] := v[j] * 2
```

However, this approach has two drawbacks. Firstly, at any one time we cannot slice two structures of the same type. In our matrix addition example we might slice *c* but, if we do, we cannot slice *a* or *b*. The procedure would be as in Fig. 1.15. Secondly, this slicing may have no run-time effect. The with-statement is essentially a *notational* convenience; it need have no effect upon the machine code generated by a compiler.

Consequently, we shall not introduce records solely to facilitate slicing; we shall use them only when the application suggests that they are appropriate, as suggested in Section 1.3.3 (p.14).

```
procedure AddMats (a, b : rowvecs;   var c : rowvecs);
  var
    i, j : subs;
begin
  for i := 1 to n do
  with c[i] do
    for j := 1 to n do
      v[j] := a[i].v[j] + b[i].v[j]
end { Add mats };
```

Figure 1.15

1.6.3 Routine slicing

A formal parameter of a routine is local to that routine and we reduce subscripting if we pass a slice of an array as a parameter. If we use a value parameter we get a local copy and, if we use a variable parameter, the slice is not copied but we get a local reference to the slice. Thus use of a value parameter gives a slicing technique equivalent to constructing a local copy as in Section 1.6.1 but use of a variable parameter gives a facility we cannot achieve by other means. In the procedure of Fig. 1.16, the handling of *a* and *b* is as in Fig. 1.12 except that the copying of rows is performed implicitly by the value-parameter mechanism rather than by explicit assignments. However, the handling of c is improved. No local copy is constructed; the variable-parameter mechanism gives a local reference to the slice.

```
procedure AddMats (a, b : nbyn;   var c : nbyn);

  var
    i : subs;

  procedure AddRows (ai, bi : vectors;   var ci : vectors);
    var
      j : subs;
  begin
    for j := 1 to n do
      ci[j] := ai[j] + bi[j]
  end { Add rows };

begin { Add mats }
  for i := 1 to n do
    AddRows (a[i], b[i], c[i])
end { Add mats };
```

Figure 1.16

This is often the most appealing approach to slicing because it encourages us to identify logically distinct activities within the overall process. For example, the matrix multiplication process defined by the procedure of Fig. 1.14 involves computation of the dot product of two vectors. This fact becomes explicit when routine slicing is used, as in Fig. 1.17.

```
procedure MultiplyMats (a, b : nbyn;   var c : nbyn);

  var
    i, j : subs;

  procedure Transpose (var m : nbyn);
    { as in Fig. 1.14 }
  . . .

  function dotprod (u, v : vectors) : real;
    var
      k : subs;
      dot : real;
```

```
    begin
      dot := 0;
      for k := 1 to n do
        dot := dot + u[k] * v[k];
      dotprod := dot
    end { dotprod };

begin { Multiply mats }
  Transpose (b);
  for i := 1 to n do
    for j := 1 to n do
      c[i,j] := dotprod (a[i], b[j])
end { Multiply mats };
```

Figure 1.17

Chapter 2 ROUNDING ERRORS

A feature of numerical methods is that they furnish only approximate solutions; a *deliberate* error is made (the truncation of a series, perhaps) so that the problem can be reformulated as one suitable for computer solution. The study of this deliberate error, and its effects, is a proper and important one but does not give the complete picture. Another type of error, with which everyone is familiar, is the *bungle*. Bungles are probably not a profitable area of study except, perhaps, to a psychologist but their existence must never be forgotten; when no other reason can be found for erroneous results then the search for a bungle should be started.

2.1 Round-off errors

A third type of error, that discussed in this chapter, is classified under the general heading of *round-off* errors. Round-off errors are a direct result of the limitations of computers; the arithmetic in a machine involves values with only a finite number of digits so that when values are combined through an arithmetic operation, errors are automatically made. Numbers are stored in the computer as a sequence of binary digits, or *bits*, but, in order to analyse the effects of round-off errors, it is assumed that numbers are represented in the *normalised* decimal *floating-point* form

$$\pm 0.d_1d_2...d_m \times 10^n \qquad (2.1)$$
$$1 \leq d_1 \leq 9, \quad 0 \leq d_i \leq 9 \qquad \text{for } i = 2,3,...,m$$

The sequence of digits $d_1d_2...$ in (2.1) is known as the *mantissa* and the index, n, as the *exponent*. As an example, the fraction $-\frac{5}{8}$ has the form

$$-0.625 \times 10^0$$

and the value $\frac{25}{4}$ has the form

$$+0.625 \times 10^1$$

The finite size of the computer implies that there is a maximum number, k say, of digits by which a value can be represented; that is, the mantissa must contain only k digits. Any real number, x, can be written in the form

$$x = \pm 0.d_1d_2...d_kd_{k+1}... \times 10^n$$

The floating point form (2.1), denoted by $fl(x)$, is obtained by terminating the mantissa of x after k digits. There are two common ways of doing this. One is known as *chopping* or *truncating*; the digits $d_{k+1}, d_{k+2}, \ldots$ are chopped from the mantissa to give

$$fl(x) = \pm 0.d_1 d_2 \ldots d_k \times 10^n$$

The second method is known as *rounding*. If x is positive then $5 \times 10^{n-(k+1)}$ is added to the number which is then chopped to give

$$fl(x) = \pm 0.D_1 D_2 \ldots D_k \times 10^n$$

If x is negative then $5 \times 10^{n-(k+1)}$ is subtracted from the number before it is chopped. For example, if k=4 and rounding is used, then 1.23456 and −6.54321 are represented as $+0.1235 \times 10^1$ and -0.6543×10^1 respectively. If k=4 and chopping is used then the representations are $+0.1234 \times 10^1$ and -0.6543×10^1.

There are also limitations on the size of the exponent; n must satisfy the inequality

$$-m \leq n \leq M$$

where M and m are positive integers which may differ for different machines. If n becomes larger than M, then the number is said to have *overflowed*; it has become too large to be represented in the machine. Overflow usually occurs when an attempt is made to divide by a very small number, usually zero, and the programmer should be on the lookout for situations where this is likely to occur. *Underflow* occurs when n becomes smaller than $-m$; in this case some computers reset the value of the number to zero and continue the calculation, others give an error message.

During a computation, the accumulation of round-off errors can completely swamp the solution so that it is essential to be able to identify those operations which can lead to large round-off errors. Two measures are usually used to quantify these errors.

Definition 2.1

If x^* is an approximation to x, then the *absolute* error is given by $|x-x^*|$ and the *relative* error is given by $|x-x^*|/|x|$, provided that x is not zero.

From this definition we see that the floating-point representation of x has a relative error

$$\frac{|x - fl(x)|}{|x|}$$

If k decimal digits are available, then a relative error bound of 10^{-k+1} is found for chopping and 5×10^{-k} for rounding.

Definition 2.2

The numbers x and x^* are said to agree to s *significant* digits (or figures) if s is the largest non-negative integer for which

$$\frac{|x-x^*|}{|x|} < 5 \times 10^{-s}$$

For example, the two numbers 123.45 and 124.35 agree to two significant figures, as do 0.012345 and 0.012435. The concept of two numbers agreeing to a certain number of significant figures is particularly useful whenever the numbers are large or small. When they are large, then the number of decimal places of agreement is probably not important; similarly, when they are small then the number of leading zero decimal places is not helpful in deciding whether or not the numbers are 'close'. For this reason, the relative error is normally used to define the closeness of numbers; it corresponds to the intuitive test of closeness that most people automatically apply when they first look at numbers.

What happens to round-off errors when numbers are combined? This is not an easy question to answer; the analysis can be complicated and can vary from machine to machine. We shall try to give a flavour of some of the problems that can arise by looking at specific examples. Those readers that are interested in the detailed analysis of round-off errors should consult Wilkinson (1963). Let us assume that $fl(x)$ and $fl(y)$ are the floating-point representations of x and y and that $\oplus$, $\ominus$, $\otimes$ and $\oslash$ represent the machine operations of addition, subtraction, multiplication and division. These operations can be defined as follows

$$x \oplus y = fl(fl(x) + fl(y))$$

$$x \ominus y = fl(fl(x) - fl(y))$$

$$x \otimes y = fl(fl(x) \times fl(y))$$

$$x \oslash y = fl(fl(x) \,/\, fl(y))$$

So let us now examine what happens to the two values x=5/6 and y=1/7 in 4-digit floating-point arithmetic. Table 2.1 lists the effects of the above operations on $fl(x)$=0.8333 and $fl(y)$=0.1429 using rounding. The notation, $\pm 0.d_1d_2(n)$, used in the rest of this chapter is a shorthand notation for $\pm 0.d_1d_2 \times 10^n$.

Since the largest relative error is 0.44(−3) these operations have all given satisfactory 4-digit results. To illustrate some problems that can arise we

Table 2.1

Operation	*Result*	*Correct value*	*Absolute error*	*Relative error*
$x \oplus y$	0.9762	41/42	0.95(−5)	0.98(−5)
$x \ominus y$	0.6904	29/42	0.76(−4)	0.11(−3)
$x \otimes y$	0.1191	5/42	0.52(−4)	0.44(−3)
$x \oslash y$	5.8310	35/6	0.23(−2)	0.40(−3)

consider operations involving the numbers u=0.14263 {$fl(u)$=0.1426}, v=543.21 {$fl(v)$=0.5432(+3)} and w=0.33333(−4) {$fl(w)$=0.3333(−4)}. The results of several operations involving these numbers are presented in Table 2.2.

Table 2.2

Operation	*Result*	*Correct value*	*Absolute error*	*Relative error*
$y \ominus u$	0.3000(−3)	0.2300(−3)	0.7000(−4)	0.3043(+0)
$(y \ominus u) \oslash w$	0.9001(+1)	0.6900(+1)	0.2101(+1)	0.3045(+0)
$(y \ominus u) \otimes v$	0.1630(+0)	0.1249(+0)	0.3810(−1)	0.3050(+0)
$u \oplus v$	0.5433(+3)	0.5434(+3)	0.1000(+0)	0.1840(−3)

The subtraction $y \ominus u$ results in a small absolute but a large relative error. Division by the small number w or multiplication by the large number v magnifies the absolute error without changing the relative error. Finally, the addition of the small numbers u and v produces a large absolute but a small relative error. These results show that, if significant figures are not to be lost, the arithmetic should be ordered to avoid subtracting numbers that are nearly equal, dividing by very small numbers and multiplying by very large numbers.

As an illustration of the technique of reordering arithmetic to avoid the loss of significant figures we consider the problem of determining the zeros of the polynomial

$$p(x) = x^2 - 10.1x + 1$$

using a machine carrying only 4 digits and using the well-known formula for the roots of a quadratic equation:

$$x = \frac{-b \pm (b^2-4ac)^{1/2}}{2a}$$

Assuming that $b^2-4ac>0$ and $b<0$, as in this example, the smaller (in modulus) of the two roots is given by

$$x_s = \frac{-b - (b^2-4ac)^{1/2}}{2a} \tag{2.2}$$

Now if $4ac$ is small compared with b^2 then $(b^2-4ac)^{1/2}$ is approximately equal to b and the numerator in (2.2) involves the subtraction of nearly equal quantities. To see the effect of this, we perform the calculation

$$\begin{aligned}
b^2 &= 0.1020(+3)\\
4ac &= 0.4000(+1)\\
b^2-4ac &= 0.9800(+2)\\
(b^2-4ac)^{1/2} &= 0.9899(+1)
\end{aligned}$$

giving

$$x_s = 0.1003(+0)$$

whereas the correct value is $x_s=0.1000(+0)$. This error can be avoided by calculating the larger root first, using the formula

$$x_l = \frac{-b + (b^2-4ac)^{1/2}}{2a}$$

In this case the value of x_l is $0.1000(+2)$. We now find the value of x_s by using the equality

$$x_s = \frac{c}{x_l} = \frac{1}{0.1000(+2)} = 0.1000(+0)$$

Now, despite round-off errors, both the zeros are found exactly.

In this example the problem was caused by the subtraction of numbers with significantly differing magnitudes and a comparable problem arises with addition. The summation of series is a process which can be similarly affected. When all the terms in a series have the same sign they should be accumulated in ascending order of magnitude. The sum of an alternating series is best computed as the difference between the sums of two series, one containing the positive terms and the other the negative terms. See Exercises 3 and 4 for examples of this.

2.2 Conditioning and stability

The *condition* of a function is a measure of the sensitivity of that function to small changes in its parameters. A function is *well-conditioned* if small changes in the parameters induce only a small change in the behaviour of the function, and it is *ill-conditioned* otherwise. The concept of conditioning is considered in more detail in Chapter 4 with reference to the solution of a system of linear algebraic equations. Here, ill-conditioning is illustrated by considering the classical problem, discussed by Wilkinson (1959), of finding the zeros of the polynomial

$$p(x) = (x-1)(x-2)...(x-19)(x-20)$$

of degree 20 with zeros at $x=1, 2, \ldots, 20$. If the factors of this polynomial are multiplied out and the coefficient of the power x^{19} is changed by approximately 10^{-7} then the behaviour of the polynomial changes radically. Now the polynomial has 5 pairs of complex roots and one of the real roots becomes -20.847.

The *stability* of a numerical process is an idea related to conditioning. A numerical method is said to be *stable* if small changes, including round-off errors, in the data induce only small changes in the solution of the process; the process is *unstable* otherwise. It is obviously desirable to use stable numerical methods in which the error in the numerical values does not grow, out of control.

If the error, E_n, at the nth stage of a numerical process has the form $|E_n| \propto n$, then the growth of the error is said to be *linear*. If $|E_n| \propto k^n$, for some $k>1$, then the growth of the error is said to be *exponential*; if $k<1$ the error decreases exponentially.

Normally it is difficult to avoid linear error growth; a convergent iterative process, such as that outlined in Section 1.3.3 (p.9), is one of the few in which there is a built-in control of the error. For $k>1$, k^n grows quickly, even for quite small values of n, so that exponential error growth should be avoided. A method with linear error growth is usually deemed to be stable and one with exponential error growth unstable. The following is an example of an unstable numerical process.

The values 1, 1/6, 1/36, 1/216,. . .,$1/6^n$,. . . can be generated by defining $x_0=1$, $x_1=1/6$ and using the relation

$$x_n = 37x_{n-1}/6 - x_{n-2} \tag{2.3}$$

for $n=2,3,\ldots$. The values in Table 2.3 were calculated using 4-digit arithmetic with rounding and, as can be seen, the errors are increasing rapidly; by the stage $n=4$ the solution, x_4, bears little resemblance to the correct answer.

Table 2.3

n	x_n	*Correct value*
0	1.0000	1.0000
1	0.1667	0.1667
2	0.0280	0.0278
3	0.0060	0.0046
4	0.0090	0.0008
5	0.0495	0.0001

The general solution of (2.3) is

$$x_n = c_1(1/6)^n + c_2(6)^n$$

and in order for x_0 to be 1 and x_1 to be 1/6, c_1 must take the value 1 and c_2 the value 0. But, since 4-digit representations are being used, x_0=0.1000(+1) and x_1=0.1667(+0) and so c_1=0.1000(+1) and c_2=0.5556(−5). Consequently x_n now has the form

$$x_n = 0.1000(+1) \times [0.1667(+0)]^n + 0.5556(-5) \times [0.6000(+1)]^n$$

so that an error $0.5556(-5) \times [0.6000(+1)]^n$ is being made at every stage. This is exponential error growth.

2.3 Exercises

1 Write and run a program to

(i) calculate the value x=1/1000 and add x to itself 999 times,
(ii) add the value 0.001 to itself 999 times and
(iii) multiply 0.001 by 1000.

Compare the results of the three computations.

2 Write a program to find the roots of the quadratic equation

$$x^2 - 1000.001x + 1 = 0$$

using the usual formula. {Note that the exact roots are 1000 and 0.001.}

3 Compare the output from two programs which form the sum

$$\sum_{i=1}^{n} \frac{1}{i!}$$

for several large values of n. One program is to accumulate terms in ascending order of magnitude and the other in descending order.

4 Write two programs to approximate e^{-x} by summing the first n terms of the series

$$\sum_{i=0}^{\infty} (-1)^i \frac{x^i}{i!}$$

The first is to sum the series directly and the other is to split the series to avoid alternating signs. Use these programs to approximate $e^{-0.1}$ and compare the results.

5 The sequence of values 1, 1/6, 1/36, 1/216,. . . can be generated by setting $x_0=1$ and defining

$$x_{n+1} = x_n/6$$

Using 4-digit rounded arithmetic calculate $x_1,. . .,x_5$ and compare them with the results in Table 2.3. Show that this algorithm exhibits linear error growth.

Chapter 3 NON-LINEAR ALGEBRAIC EQUATIONS

The problem of finding a value α which is a zero of a function $f(x)$ and hence a root of the equation

$$f(x) = 0$$

occurs frequently. For example, whenever the maximum of a function $F(x)$ is sought a necessary condition which must be satisfied at the maximum is that the derivative of $F(x)$ should be zero. In practice this condition is used as an aid to finding the maximum. The techniques discussed in this chapter, along with those for solving systems of linear algebraic equations, underpin many of the numerical methods used to solve applied mathematical problems.

This chapter is split into four sections. The first concerns methods which do not use derivatives, the second introduces a method which does use derivatives and, in the third and fourth, this method is applied to polynomial functions and systems of non-linear equations. We shall assume throughout that $f(x)$ is continuous in the interval of interest and is sufficiently differentiable for the methods to be meaningful.

Throughout the first two sections the performances of the different methods on the test function

$$f(x) = x - e^{1/x}$$

will be compared. This function has only one zero in the interval [1,2] and, to six decimal places, this is 1.763223.

3.1 Methods without derivatives

Throughout this section it is assumed that the function, $f(x)$, is continuous and that an interval containing the root has already been obtained using, for example, a rough graph of the function.

3.1.1 Bisection

In the method of bisection, an interval containing the root is successively halved until the size of the interval reaches some specified tolerance. Given that there is only one root in the interval $[x_0,x_1]$, say, then

$$f(x_0)f(x_1) \leq 0$$

with equality holding if either x_0 or x_1 is the root. Fig. 3.1 illustrates this.

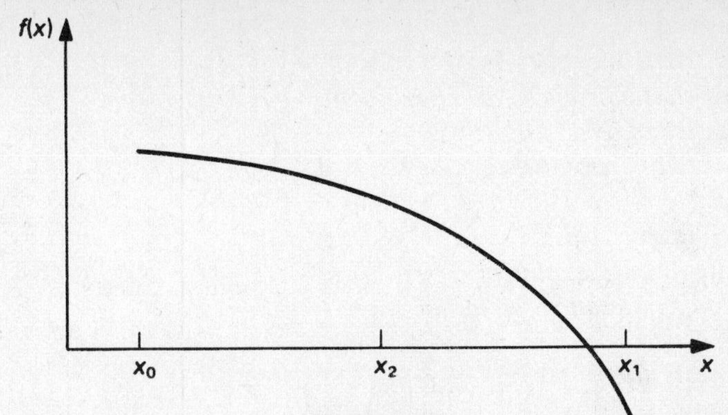

Figure 3.1

The mid-point of the interval, x_2, is determined, viz.

$$x_2 = (x_0 + x_1)/2$$

and the function value, $f(x_2)$, calculated. If

$$f(x_0)f(x_2) \leq 0$$

then the root must lie in the interval $[x_0,x_2]$, otherwise it lies in the interval $[x_2,x_1]$. In Fig. 3.1 the root lies in $[x_2,x_1]$. In either case the interval in which the root lies has been halved or bisected and further bisection can be carried out on the smaller interval until the remaining interval is of the required size. Clearly, after n steps of bisection, the remaining interval has size

$$\frac{x_1 - x_0}{2^n}$$

When the size of the interval containing the root does not exceed some value t, any value within that interval differs from the root by no more than t. By specifying t we are specifying the error we are prepared to tolerate in the value of the root.

The procedure of Fig. 3.2 applies the method. The midpoint of the final interval is returned via the variable parameter *root* giving a maximum error in the approximation to the root of *xtol*/2. The function value at this point is returned via *fatroot*. The initial interval is specified by two value parameters *x0* and *x1*. Because they are value parameters, changes to their values within the procedure do not affect any variables supplied as actual parameters when the procedure is called. The variable parameter *noofits* returns the number of iterations performed and *converged* indicates whether the process converged within the maximum number of iterations allowed (*maxits*). The subrange type *posints* is

1 . . *maxint*

```
procedure Bisect
   (function f (x : real) : real;
    x0, x1, xtol : real;    maxits : posints;
    var root, fatroot : real;    var noofits : posints;
    var converged : boolean);
      { This procedure assumes x0 < x1 }
   var
      x, fx, f0 : real;
      itcount : 0 . . maxint;
      state : (iterating, withintol, maxitsreached);
begin
   f0 := f(x0);
   itcount := 0;    state := iterating;
   repeat
      itcount := itcount + 1;
      x := (x0 + x1) / 2;    fx := f(x);
      if f0 * fx <= 0 then x1 := x else
      begin
         x0 := x;    f0 := fx
      end;
      if x1-x0 <= xtol then state := withintol else
         if itcount = maxits then state := maxitsreached
   until state <> iterating;
   converged := state = withintol;
   root := (x0 + x1)/2;    fatroot := f(root);
   noofits := itcount
end { Bisect };
```

Figure 3.2

as defined in Section 1.3.

For our test function, a sample call might be

Bisect (*f*, 1, 2, *tol*, 50, *x*, *fx*, *its*, *success*)

within the environment

```
const
   tol = 5E-7;

type
   posints = 1 . . maxint;
```

```
var
   x, fx : real;
   its : posints;
   success : boolean;

function f (x : real) : real;
begin
   f := x - exp (1/x)
end { f };
```

In practice the program should be run at least twice, using a different value of *tol* each time, and a check made that the values of x (and fx) agree.

3.1.2 Regula falsi

Although the bisection method is guaranteed to converge it may do so only slowly. It is also inefficient in the sense that it is not using all the available information; it takes note of the signs of f_0, f_1 and f_2 but not of their values.

Consider the function represented in Fig. 3.1. The bisection method always chooses the mid-point of the interval as the next point, regardless of the fact that f_1 is small indicating that the root may be near x_1. The *regula falsi* method, sometimes known as the method of *false position*, takes the function values into account in a particularly simple way; the straight line joining f_0 and f_1 is drawn and the point where this line cuts the x-axis is defined to be the new point, x_2.

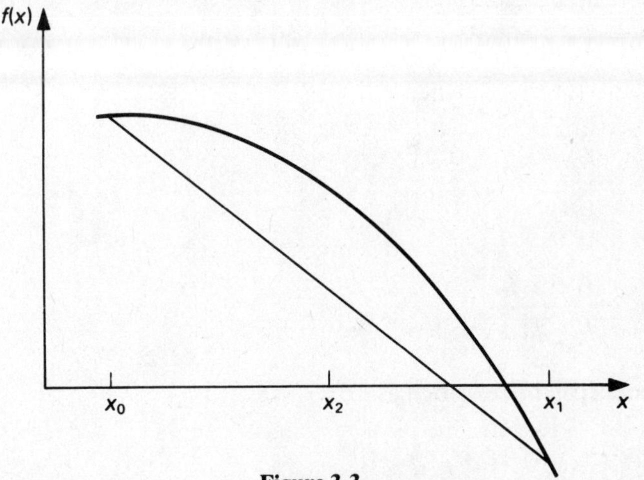

Figure 3.3

As can be seen in Fig. 3.3 the point x_2, for this particular function, is a better approximation to the zero than is the mid-point and also the remaining interval is smaller. This is not the case for every function and Fig. 3.4 shows a function for which the size of the interval in the regula falsi method decreases more slowly than that in the bisection method.

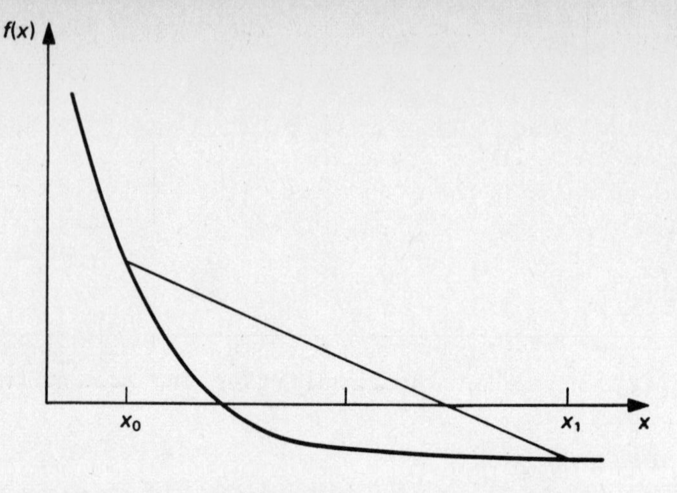

Figure 3.4

In practice the line joining f_0 and f_1 is not drawn; instead, the value of x_2 is calculated directly. The equation of the line joining the two points can be written in the form

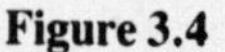

$$y = \frac{(x - x_1)}{x_0 - x_1} f_0 + \frac{(x - x_0)}{x_1 - x_0} f_1$$

$$= \frac{x(f_1 - f_0) - x_0 f_1 + x_1 f_0}{x_1 - x_0}$$

which is zero (at $x = x_2$) when

$$x_2(f_1 - f_0) = x_0 f_1 - x_1 f_0$$

that is

$$x_2 = \frac{x_0 f_1 - x_1 f_0}{f_1 - f_0} \tag{3.1}$$

This formula can be rewritten as

$$x_2 = x_1 - \frac{(x_1 - x_0)}{f_1 - f_0} f_1 \tag{3.2}$$

Intuitively, this form is more appealing because the correction to x_1 required to obtain x_2 appears explicitly.

Having calculated x_2, the process applied to determine the remaining interval is as in the bisection method. The regula falsi method suffers from the disadvantage that, in general, the size of the interval containing the root does not tend to zero. This is because, in the neighbourhood of the root, most functions are strictly concave or convex and this results in one of the end points approaching the root whilst the other remains fixed. Consequently, the size of the interval is of little use as a bound on the error in the approximation or as a tolerance for the termination of the iteration. The secant method, a modification of the regula falsi method, overcomes this drawback.

3.1.3 Secant

The secant method removes the restriction that successive iterates must

```
procedure Secant
  (function f (x : real) : real;
   x0, x1, xtol : real;   maxits : posints;
   var root, fatroot : real;   var noofits : posints;
   var outcome : secoutcomes);
  const
    assumedzero = 1E−20;
  var
    f0, f1, x2 : real;
    itcount : 0 . . maxint;
    state : secstates;
begin
  f0 := f(x0);   f1 := f(x1);
  itcount := 0;   state := iterating;
  repeat
    if abs(x1) <= assumedzero then state := toonearzero else
      if abs(f1−f0) <= assumedzero then state := tooflat else
      begin
        itcount := itcount + 1;
        x2 := x1 − (x1 − x0) * f1/(f1 − f0);
        if abs( (x2 − x1)/x1) <= xtol then state := withintol else
          if itcount = maxits then state := maxitsreached else
          begin
            x0 := x1;   x1 := x2;
            f0 := f1;   f1 := f(x1)
          end
      end
  until state <> iterating;
  outcome := state;
  root := x2;   fatroot := f(x2);   noofits := itcount
end { Secant };
```

Figure 3.5

bracket the root. Successive function values may now have the same sign and a consequence of this is that the iteration process becomes more sensitive to rounding errors. However, the effects of this are less when formula (3.2) is used in preference to (3.1).

The procedure of Fig. 3.5 implements the method directly and applies, to successive iterates, the usual *relative* error test described in Chapter 2. As with the bisection method, successive iterates may converge or the maximum permitted iterations may be reached but now, in addition, there are two checks for overflow that should be made. Overflow can result from division either by $f_1 - f_0$ or by x_1. Consequently, the boolean parameter *converged* of Fig. 3.2 has been replaced by a parameter *outcome* of symbolic type because there are several reasons for terminating the iteration. The following Pascal environment is assumed.

```
type
  posints = 1 . . maxint;
  secstates = (iterating, withintol,
                    maxitsreached, toonearzero, tooflat);
  secoutcomes = withintol . . tooflat;
```

The procedure of Fig. 3.6 expresses the algorithm differently and is probably marginally more efficient as a result. No longer is x_1 always updated to identify the latest estimate and x_0 the previous value; instead their roles are successively interchanged. A boolean variable *x0iscurrent* indicates whether or not *x0* is the current estimate. A slight gain in speed may result because updating and checking a boolean variable is faster than updating two real variables. However, the difference will be so slight as to be insignificant; you should adopt the approach you find more transparent.

```
procedure Secant
  (function f (x : real) : real;
   x0, x1, xtol : real;    maxits : posints;
   var root, fatroot : real;    var noofits : posints;
   var converged : secoutcomes);
  const
    assumedzero = 1E-20;
  var
    f0, f1, newf : real;
    itcount : 0 . . maxint;
    x0iscurrent : boolean;
    state : secstates;
begin
  f0 := f(x0);   f1 := f(x1);   x0iscurrent := false;
  itcount := 0;   state := iterating;
```

```
repeat
  if abs(f1 - f0) <= assumedzero then state := tooflat else
  begin
    itcount := itcount + 1;
    if x0iscurrent then
    begin
      x1 := x1 - (x1 - x0) * f1/(f1 - f0);
      f1 := f(x1);   newf := f1
    end else
    begin
      x0 := x0 - (x0 - x1) * f0/(f0 - f1);
      f0 := f(x0);   newf := f0
    end;
    x0iscurrent := not x0iscurrent;
    if abs(x0) <= assumedzero then state := toonearzero else
      if abs( (x1-x0)/x0) <= xtol then state := withintol else
        if itcount = maxits then state := maxitsreached
  end
until state <> iterating;
outcome := state;
if x0iscurrent then
begin
  root := x0;   fatroot := f0
end else
begin
  root := x1;   fatroot := f1
end;
noofits := itcount
end { Secant };
```

Figure 3.6

The values in Table 3.1 are the first five iterates produced by the procedures of Figs. 3.2 and 3.5 (or 3.6) when applied to our test function with

Table 3.1

Iteration	*Bisect*	*Secant*
1	1.500000	1.830264
2	1.750000	1.759565
3	1.875000	1.763286
4	1.812500	1.763223
5	1.781250	1.763223
Convergence	21	5

initial interval [1,2]. The figure at the foot of each column indicates how many iterations are required for convergence when *xtol* is 5E−7. Unfortunately, in certain circumstances, the secant method may not converge at all and this possibility will be covered in Section 3.2 when we discuss Newton's method.

3.1.4 Fixed-point iterations

A different class of methods, known as fixed-point iterations, can be generated by taking account of the form of the function. Although some of these methods use more than one approximation, each new value is usually calculated using only the previous approximation to the root.

The iteration giving the new value generally has the form

$$x_{n+1} = g(x_n), \qquad n = 0,1,2,\ldots \tag{3.3}$$

and, using any information about the position of the root, an initial guess, x_0, at the value of the root must be provided. The equation

$$x = g(x)$$

is obtained by rearranging the equation

$$f(x) = 0$$

so that an x is obtained on the left-hand side. For example, the equation

$$x - e^{1/x} = 0$$

can be rearranged to give

$$x = e^{1/x} \tag{3.4}$$

so that in this case

$$g(x) = e^{1/x} \tag{3.5}$$

Different functions $g(x)$ can be obtained by taking alternative forms of equation (3.4). Taking natural logarithms of both sides we have

$$\ln(x) = 1/x$$

which gives

$$g(x) = 1/\ln(x) \tag{3.6}$$

Adding x to both sides of (3.4) we obtain

$$2x = x + e^{1/x}$$

and this gives

$$g(x) = \frac{x + e^{1/x}}{2} \tag{3.7}$$

The method proceeds, given $g(x)$ and x_0, by calculating

$$x_1 = g(x_0)$$

then

$$x_2 = g(x_1)$$

and so on.

Definition 3.1

If $g(x)$ is defined on the interval $[a,b]$ and $g(\alpha) = \alpha$ for some α in $[a,b]$, then the function $g(x)$ is said to have a *fixed point* α in $[a,b]$.

Theorem 3.1

If $g(x)$ is continuous on $[a,b]$ and $g(x)$ is in $[a,b]$ for all x in $[a,b]$, then $g(x)$ has a fixed point in $[a,b]$. Further, suppose that $g'(x)$ exists on (a,b) and

$$| g'(x) | \leq K < 1 \tag{3.8}$$

for all x in (a,b), then g has a unique fixed point α in $[a,b]$.

Proof

If $g(a) = a$ (or $g(b) = b$), then a (or b) is obviously a fixed point. Suppose that this is not the case; then it must be true that $g(a) > a$ and $g(b) < b$, otherwise $g(x)$ is not in $[a,b]$. Define $h(x) = g(x) - x$; then h is continuous on $[a,b]$ with

$$h(a) = g(a) - a > 0 \qquad \text{and} \qquad h(b) = g(b) - b < 0$$

The intermediate value theorem (Rudin, 1964, p.81) implies that there exists a point α in (a,b) for which $h(\alpha) = 0$. Thus $g(\alpha) = \alpha$ and α is a fixed point of g.

Suppose also that

$$| g'(x) | \leq K < 1$$

for all x in $[a,b]$ and that α and β are both fixed points in $[a,b]$ with $\alpha \neq \beta$. By the mean value theorem (Rudin, 1964, p.93) there exists a point ξ in $[a,b]$ such that

$$\begin{aligned} |\alpha - \beta| &= |g(\alpha) - g(\beta)| = |g'(\xi)|\ |\alpha - \beta| \\ &\leq K|\alpha - \beta| < |\alpha - \beta| \end{aligned}$$

which is a contradiction. Hence $\alpha = \beta$ and the fixed point in $[a,b]$ is unique. ■

The conditions set in Theorem 3.1 can be used to confirm that a function $g(x)$ has a fixed point α in some interval. The conditions are, however, only sufficient and not necessary. Consider, for example, the function $g(x) = e^{-2x}$ on $[0,1]$. The first derivative of $g(x)$ is

$$g'(x) = -2e^{-2x}$$

and, at the point $x=0.1$, takes the value -1.6375, correct to four decimal places. Nevertheless, e^{-2x} is a strictly decreasing function on $[0,1]$ and x is strictly increasing and a graph will show that there is a single fixed point.

The following theorem provides a test which can be used to check whether iteration (3.3) gives iterates which converge to the fixed point, the zero of our original function $f(x)$.

Theorem 3.2

Let $g(x)$ be continuous on $[a,b]$ and $g(x)$ be in $[a,b]$ for all x in $[a,b]$. If $g'(x)$ exists on (a,b) and

$$|g'(x)| \leq K < 1$$

for all x in (a,b) then, if x_0 is any value in $[a,b]$, the sequence

$$x_{n+1} = g(x_n), \qquad n \geq 0$$

converges to the unique fixed point, α, in $[a,b]$.

Proof

The function $g(x)$ maps $[a,b]$ onto itself so that the sequence x_n, $n = 0,1,2. . .$, is well defined and its values lie in $[a,b]$. Define the error in the iterate x_n to be

$$\epsilon_n = x_n - \alpha$$

Then using the mean value theorem and inequality (3.8)

$$\begin{aligned} |\epsilon_n| &= |x_n - \alpha| = |g(x_{n-1}) - g(\alpha)| \\ &= |g'(\xi)|\ |x_{n-1} - \alpha| = |g'(\xi)|\ |\epsilon_{n-1}| \\ &\leq K|\epsilon_{n-1}| \end{aligned} \tag{3.9}$$

where ξ is in $[a,b]$. Applying this argument again,

$$
\begin{aligned}
|\epsilon_n| &\le K|\epsilon_{n-1}| \le K.K|\epsilon_{n-2}| \\
&= K^2|\epsilon_{n-2}| \le \cdots \le K^n|\epsilon_0|
\end{aligned}
$$

Now since $K < 1$

$$\lim_{n\to\infty} |\epsilon_n| = \lim_{n\to\infty} K^n |\epsilon_0| = 0$$

and the sequence converges to α. ■

Definition 3.2

Suppose that x_n, $n=0,1,2,\ldots$, is a sequence that converges to α and let $\epsilon_n = x_n - \alpha$ for each n. If positive constants c and p exist with

$$\lim_{n\to\infty} \frac{|\epsilon_{n+1}|}{|\epsilon_n|^p} = c$$

then the sequence of iterates, x_n, is said to converge to α with order p and asymptotic error constant c.

If the function $g(x)$ has a fixed point α, then the iteration (3.3)

$$x_{n+1} = g(x_n)$$

is said to be of order p if the sequence of iterates x_n, $n=0,1,2,\ldots$, is convergent to α with order p. By considering Equation 3.9 it can be shown that for a function $g(x)$ satisfying the conditions of Theorem 3.2, the sequence of iterates produced converges to α with order at least 1.

We now consider the application of Theorem 3.2 to each of the three rearrangements (3.5), (3.6) and (3.7) of our test function. We will not check the behaviour of $g(x)$ throughout an interval, as in the theorem, but will check its derivative value at the initial guess only; if this value is strictly less than 1 then, provided that the initial guess is sufficiently close to the root, convergence is guaranteed. The mid-point, $x_0 = 1.5$, of the given interval is chosen as our initial guess. Checking the condition (3.8) for (3.5)

$$g(x) = e^{1/x}$$

gives

$$g'(x) = -\frac{e^{1/x}}{x^2}$$

so that

$$|g'(x_0)| = 0.8657 < 1$$

and the iterates will converge. When applied to (3.6)

$$g(x) = 1/\ln(x)$$

then

$$g'(x) = -\frac{1}{x(\ln(x))^2}$$

and

$$|g'(x_0)| = 4.0551 > 1$$

so the iterates will diverge. For (3.7)

```
procedure FixedPoint
  (function g (x : real) : real;
   x0, xtol : real;   maxits : posints;
   var root : real;   var noofits : posints;
   var outcome : fpoutcomes);
  const
    assumedzero = 1E-20;
  var
    x, oldx : real;
    itcount : 0 . . maxint;
    state : fpstates;

begin
  x := x0;   itcount := 0;   state := iterating;
  repeat
    if abs(x) <= assumedzero then state := toonearzero else
    begin
      itcount := itcount + 1;   oldx := x;
      x := g(x);
      if abs( (x-oldx)/oldx) <= xtol then state := withintol else
        if itcount = maxits then state := maxitsreached
    end
  until state <> iterating;
  outcome := state;
  root := x;   noofits := itcount
end { Fixed point };
```

Figure 3.7

$$g(x) = \frac{x + e^{1/x}}{2}$$

$$g'(x) = 0.5 - \frac{e^{1/x}}{2x^2}$$

and

$$|\, g'(x_0) \,| = 0.0672 < 1$$

giving a convergent sequence of iterates.

It is important that this test should be carried out before the program to calculate the iterates is run. Iteration (3.6) gives a divergent sequence of iterates and so the solution would never be found even if the program were to run forever.

The procedure of Fig. 3.7 implements a fixed-point iteration and assumes the following Pascal environment.

```
type
  fpstates = (iterating, withintol, maxitsreached, toonearzero);
  fpoutcomes = withintol . . toonearzero;
```

The process terminates when, using the relative error check, two successive values of x agree to within some specified tolerance. As usual, the procedure accepts a maximum number of iterations so the program will not run forever should the wrong iteration formula be included by mistake and a check is made for attempted division by zero. Because the iteration makes no use of the original function f, the value of the function at the computed root is not directly available.

Table 3.2 contains the first few iterates produced by iterations (3.5), (3.6) and (3.7) when supplied with the initial value 1.5. The number of steps

Table 3.2

Iteration	(3.5)	(3.6)	(3.7)
1	1.947734	2.466303	1.723867
2	1.670991	1.107763	1.755034
3	1.819291	9.771126	1.761464
4	1.732672	0.438706	1.762843
5	1.780944	−1.213701	1.763141
6	1.753301	—	1.763205
Convergence	25	—	9

required for convergence are for a tolerance of 5E−7. As predicted (3.5) and (3.7) give convergent sequences whereas the sequence produced using (3.6) diverges. Only five iterates are obtained with (3.6) because, if a sixth is attempted, the function *ln* is supplied with a negative parameter and so the program fails.

It is obvious from (3.9) in Theorem 3.2 that the smaller the value of K the more rapid will be the convergence. For this reason we would expect iteration (3.7) to converge more rapidly than (3.5) and this is verified by the results presented in Table 3.2.

3.1.5 Aitken's Δ^2 process

Aitken's Δ^2 process is a well-known acceleration technique which, when used in conjunction with a first-order iteration method, can be shown (Isaacson and Keller 1966, p.102) to give second-order convergence unless there are multiple roots, in which case the order of convergence remains 1. Following (3.9), when the iteration has progressed sufficiently, a first-order method has the property that

$$\epsilon_{n+1} \simeq \epsilon_n K$$

or

$$\frac{\epsilon_{n+1}}{\epsilon_n} \simeq K$$

where $\epsilon_n = \alpha - x_n$. Similarly

$$\frac{\epsilon_{n+2}}{\epsilon_{n+1}} \simeq K$$

so that

$$\frac{\epsilon_{n+2}}{\epsilon_{n+1}} \simeq \frac{\epsilon_{n+1}}{\epsilon_n}$$

or

$$\frac{\alpha - x_{n+2}}{\alpha - x_{n+1}} \simeq \frac{\alpha - x_{n+1}}{\alpha - x_n}$$

This can be rearranged to give

$$\alpha \simeq x_n - \frac{(x_{n+1} - x_n)^2}{x_{n+2} - 2x_{n+1} + x_n}$$

```
procedure Aitken
  (function g (x : real) : real;
   x0, xtol : real;   maxits : posints;
   var root : real;   var noofits : posints;
   var outcome : fpoutcomes);
  const
    assumedzero = 1E-20;
  var
    itcount : 0 . . maxint;
    x, oldx, x1, x2 : real;
    state : fpstates;

begin
  x := x0;
  itcount := 0;   state := iterating;
  repeat
    if abs(x) <= assumedzero then state := toonearzero else
    begin
      oldx := x;   itcount := itcount + 1;
      case itcount mod 3 of
        0 : begin
              x0 := x0 - sqr(x1 - x0)/(x2 - 2 * x1+x0);
              x := x0
            end;
        1 : begin
              x1 := g(x0);   x := x1
            end;
        2 : begin
              x2 := g(x1);   x := x2
            end
      end { case };
      if abs( (x - oldx)/oldx) <= xtol then state := withintol else
        if itcount = maxits then state := maxitsreached
    end
  until state <> iterating;
  outcome := state;
  root := x;   noofits := itcount
end { Aitken };
```

Figure 3.8

This formula gives an approximation to α which can be used as the next iterate. So x_{n+3} is defined by

$$x_{n+3} = x_n - \frac{(x_{n+1} - x_n)^2}{x_{n+2} - 2x_{n+1} + x_n} \tag{3.10}$$

or in the notation of finite differences

$$x_{n+3} = x_n - \frac{(\Delta x_n)^2}{\Delta^2 x_n}$$

In general x_{n+3}, calculated from (3.10), is more accurate than if the formula

$$x_{n+3} = g(x_{n+2})$$

were used, especially if the iteration has progressed so that successive errors satisfy (3.9) more closely. The process can continue by calculating two further iterates x_{n+4} and x_{n+5} using (3.3) and then applying a further Aitken Δ^2 step, (3.10), to evaluate x_{n+6} and so on.

This acceleration technique is implemented in the procedure of Fig. 3.8. The three previous iterates required for the application of an Aitken step are stored in three separate variables. They could, instead, be stored in an array as in Fig. 3.9. The two approaches are equally good.

Table 3.3

Iteration	(3.5)	(3.7)
1	1.947734	1.723867
2	1.670991	1.755034
3	1.776704	1.760075
4	1.755651	1.762544
5	1.767541	1.763076
6	1.763250	1.763222
Convergence	10	7

The values in Table 3.3 were produced by applying Aitken's Δ^2 process to iterations (3.5) and (3.7). As usual the starting value was 1.5 and the tolerance was 5E−7. Comparison with Table 3.2 shows that the effect of the Aitken acceleration is more pronounced with (3.5) than with (3.7). This happens because iteration (3.7) is already close to being second order.

```
procedure Aitken
  (function g (x : real) : real;
   x0, xtol : real;   maxits : posints;
   var root : real;   var noofits : posints;
   var outcome : fpoutcomes);
  const
    assumedzero = 1E-20;
  var
    i, itcount : 0 . . maxint;
    oldx : real;
    x : array [0. .2] of real;
    state : fpstates;
begin
  itcount := 0;   i := itcount;
  x[0] := x0;   state := iterating;
  repeat
    if abs(x[i]) <= assumedzero then state := toonearzero else
    begin
      oldx := x[i];
      itcount := itcount + 1;   i := itcount mod 3;
      case i of
           0 : x[0] := x[0] - sqr(x[1] - x[0]) / (x[2] - 2 * x[1] + x[0]);
         1,2 : x[i] := g(x[i-1])
      end { case };
      if abs( (x[i] - oldx)/oldx) <= xtol then state := withintol else
      if itcount = maxits then state := maxitsreached
    end
  until state <> iterating;
  outcome := state;
  root := x[i];   noofits := itcount
end { Aitken };
```

Figure 3.9

3.2 Newton's method

It was mentioned earlier that the smaller the value of $| g'(\alpha) |$ or $| g'(x_0) |$ the faster the convergence of the iterates. Newton's method, sometimes known as the Newton–Raphson method, has the property that in general

$$g'(\alpha) = 0 \qquad \text{and} \qquad g''(\alpha) \neq 0$$

This means that the error in the $(n+1)$st iterate is proportional to the square of the error in the nth iterate; i.e. the method is second-order convergent.

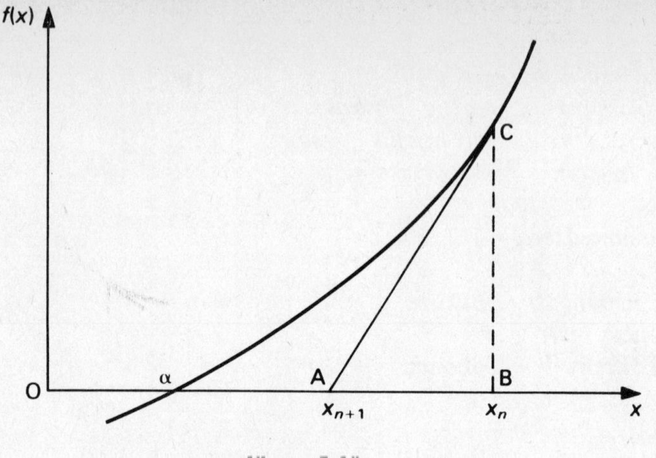

Figure 3.10

The Newton iteration can be derived by expanding $f(\alpha)$ about x_n using Taylor's series and ignoring terms of order $(\alpha - x_n)^2$ and higher or, alternatively, by geometric considerations. The latter approach gives some feeling for the way that the method works.

Consider Fig. 3.10 in which the graph of a function $f(x)$ is drawn. A root, α, of the function is indicated and the existing estimate, x_n, of the root is represented by OB. The ordinate CB is the value $f(x_n)$ and the next iterate, x_{n+1}, is chosen to be the point at which the tangent to $f(x)$ at x_n cuts the x-axis. This tangent is represented by AC and so x_{n+1} is OA and is a better estimate than x_n. Now

$$\mathrm{OA} = \mathrm{OB} - \mathrm{AB}$$

and

$$\mathrm{AB} = \frac{\mathrm{CB}}{\tan(\mathrm{CAB})}$$

where tan (CAB) is the gradient of the tangent at C and is given by $f'(x_n)$. Hence

$$x_{n+1} = \mathrm{OA} = \mathrm{OB} - \mathrm{AB} = \mathrm{OB} - \frac{\mathrm{CB}}{\tan(\mathrm{CAB})}$$

$$= x_n - f(x_n)/f'(x_n) \tag{3.11}$$

Obviously problems with this iteration can arise whenever

$$f'(x) = 0$$

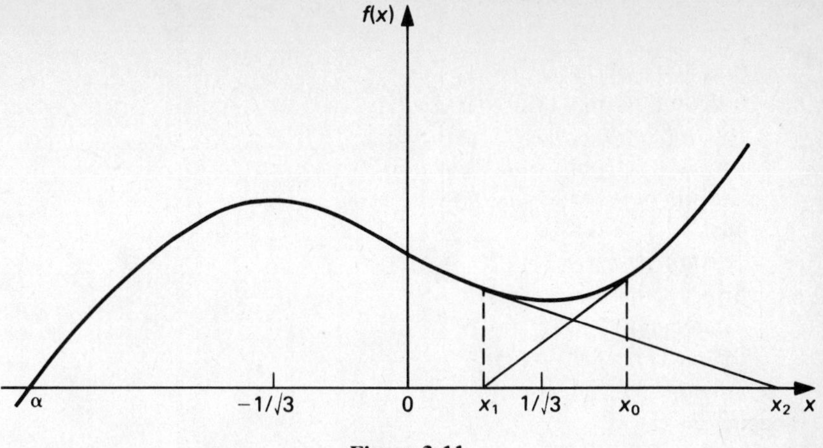

Figure 3.11

in the interval between x_0 and α where, for example, $f(x)$ has a local maximum or minimum, as in Fig. 3.11, or where $f(x)$ has a multiple root at α.

Fig. 3.11 shows a graph of $f(x) = x^3 - x + 1$. An initial guess $x_0 > -1/\sqrt{3}$ gives iterates which oscillate around $x = 1/\sqrt{3}$, at which point $f(x)$ has a local minimum. The method converges only if a subsequent iterate takes a value less than $-1/\sqrt{3}$.

The problem of maxima or minima can be overcome by attempting to sketch the function and so obtain a better initial guess. The problem of multiple roots is more complex and will not be covered here; however it is possible to show that, in this case, Newton's method still converges although only with first order. More generally the following theorem, whose proof is left as an exercise for the reader, is valid for Newton's method.

Theorem 3.3

Let α be a solution of $x = g(x)$. Suppose that $g'(\alpha) = 0$ and $g''(x)$ is continuous in an open interval containing α. Then there exists a value $d > 0$ such that, for $\alpha - d \leq x_0 \leq \alpha + d$, fixed-point iteration becomes a second-order method. Further, if $|g''(x)|/2 < M$ for $\alpha - d < x < \alpha + d$ and $M|\epsilon_0| < 1$, then the fixed-point iteration converges with order 2.

For a proof of this theorem see Burden *et al*. 1981 (p.47).

The procedure of Fig. 3.12 implements Newton's method and assumes the following Pascal environment

```
type
   posints = 1 . . maxint;
   newtstates = (iterating, withintol, maxitsreached, toonearzero, tooflat);
   newtoutcomes = withintol . . tooflat;
```

```
procedure Newton
  (function f (x : real) : real;
   function fdashed (x : real) : real;
   x0, xtol : real;   maxits : posints;
   var root, fatroot : real;   var noofits : posints;
   var outcome : newtoutcomes);
  const
    assumedzero = 1E-20;
  var
    state : newtstates;
    x, oldx, fx, fdx : real;
    itcount : 0 . . maxint;
begin
  x := x0;   fx := f(x);
  itcount := 0;   state := iterating;
  repeat
    if abs(x) <= assumedzero then state := toonearzero else
    begin
      itcount := itcount + 1;
      fdx := fdashed (x);
      if abs(fdx) <= assumedzero then state := tooflat else
      begin
        oldx := x;   x := x - fx/fdx;   fx := f(x);
        if abs( (x-oldx)/oldx) <= xtol then state := withintol else
          if itcount = maxits then state := maxitsreached
      end
    end
  until state <> iterating;
  outcome := state;
  noofits := itcount;
  root := x;   fatroot := fx
end { Newton };
```

Figure 3.12

Notice that the data types *newtstates* and *newtoutcomes* are exactly *secstates* and *secoutcomes* of Section 3.1.3.

The figures in Table 3.4 were produced when Newton's method was applied to our test function with a starting value of 1.5 and a tolerance of 5E−7. A comparison with the earlier tables shows how much more quickly Newton's method converges.

Although Newton's method generally gives extremely rapid convergence it suffers from the disadvantage that knowledge of the first derivative of $f(x)$ is required. In certain cases the calculation of this derivative can be extremely

Table 3.4

Iteration	*Newton*
1	1.739987
2	1.763078
3	1.763223
4	1.763223
Convergence	4

time consuming. In this situation the secant method should be considered because it is, in effect, Newton's method with the derivative of $f(x)$ replaced by a difference approximation, viz.

$$f'_n = \frac{f_n - f_{n-1}}{x_n - x_{n-1}}$$

Substituting this replacement into (3.11), the formula (3.2) is obtained.

The secant method and Newton's method suffer from the same problems regarding zero derivatives and multiple roots. It can be shown (Ostrowski 1973) that, in general, the secant method converges with order approximately 1.62 and so, with no necessity for a derivative evaluation, becomes very attractive compared to Newton's method.

Higher-order methods will not be considered since they usually require higher derivatives and, generally, these are difficult to obtain.

3.3 Polynomial equations

The problem of finding some or all of the zeros of the polynomial

$$p(x) = a_n x^n + a_{n-1} x^{n-1} + \cdots + a_1 x + a_0 \tag{3.12}$$

where a_n is non-zero, is one which occurs frequently. Great care must be taken when an attempt is made to solve such a problem. As has been seen in Chapter 2 difficulties can arise even when polynomials of degree 2 are involved. These difficulties multiply when polynomials of higher degree are considered.

Obviously all the methods described so far can be used to find a zero of the polynomial but the special structure of the function in this case makes it worthwhile to consider polynomials in more detail. Firstly it is known exactly how many zeros a polynomial has and, given that the coefficients a_i in (3.12) are real, it is also known that the zeros are real or occur in complex conjugate pairs. Secondly, polynomials can be evaluated extremely efficiently on a computer using a technique called nested multiplication. To evaluate (3.12) by forming x^n, multiplying by a_n and then forming x^{n-1}, etc., would involve n

additions and $n(n+1)/2$ multiplications. Even if x^n were formed as $x(x^{n-1})$, starting with $n = 2$, then n additions and $2n-1$ multiplications would be required. Nested multiplication uses only n additions and n multiplications. The polynomial (3.12) is rewritten in the form

$$p(x) = \{. . .[(a_n x + a_{n-1})x + a_{n-2}]x + \cdot \cdot \cdot + a_1\}x + a_0$$

and, for a particular value of x, is evaluated from the innermost bracket outwards; i.e. the following sequence is formed

$$\begin{aligned} p_{n-1} &= a_n \\ p_{n-2} &= p_{n-1}x + a_{n-1} \\ p_{n-3} &= p_{n-2}x + a_{n-2} \\ &\cdot \\ &\cdot \\ &\cdot \\ p_0 &= p_1 x + a_1 \\ P &= p_0 x + a_0 \end{aligned} \tag{3.13}$$

whence P is the value of $p(x)$. The function of Fig. 3.13 evaluates the polynomial $p(x)$ using the relationships (3.13). The data types *subs* and *vectors* are assumed to be as defined in Section 1.3.3 (p.13) but with the lower bound of *subs* being 0.

```
function poly (a : vectors;   n : subs;   x : real) : real;
  var
    i : subs;
    p : real;
begin
  p := a[n];
  for i := n-1 downto 0 do
    p := p * x + a[i];
  poly := p
end { poly };
```

Figure 3.13

3.3.1 Real zeros

The recurrence relation (3.13) can be derived in a different way using the method of synthetic division. Consider the problem of dividing $p(x)$ by the factor $(x-z)$. It is well known that $p(x)$ can be written in the form

$$p(x) = (x-z)q(x) + r \tag{3.14}$$

where r is the remainder and $q(x)$ is known as the quotient polynomial. Obviously

$$p(z) = r$$

The polynomial $q(x)$ has degree $(n-1)$, since $p(x)$ has degree n, and can be written in the form

$$q(x) = b_{n-1}x^{n-1} + b_{n-2}x^{n-2} + \cdots + b_1x + b_0 \tag{3.15}$$

Substituting (3.15) into (3.14) and comparing coefficients with (3.12) we get

$$\begin{aligned} a_n &= b_{n-1} \\ a_{n-1} &= b_{n-2} - zb_{n-1} \\ a_{n-2} &= b_{n-3} - zb_{n-2} \\ &\;\;\vdots \\ a_1 &= b_0 - zb_1 \\ a_0 &= r - zb_0 \end{aligned} \tag{3.16}$$

Comparison of (3.16) with (3.13) shows that each p_i, calculated in (3.13), is the coefficient b_i of the quotient polynomial $q(x)$ and P is the same as r.

It makes sense to use the fastest method discussed, Newton's method, to determine a zero of the polynomial and to do this the derivative of $p(x)$ at z is required. Now

$$p'(x) = (x-z)q'(x) + q(x)$$

so that

$$p'(z) = q(z)$$

Thus the value of the derivative of $p(x)$ at z is merely the value of the quotient polynomial, $q(x)$, at that point. Hence, recalling that $q(x)$ has the form (3.15), a further recurrence relation can be derived to give $q(z)$, viz.

$$\begin{aligned} c_{n-2} &= b_{n-1} \\ c_{n-3} &= c_{n-2}z + b_{n-2} \\ &\;\;\vdots \\ c_1 &= c_2z + b_2 \\ c_0 &= c_1z + b_1 \\ Q &= c_0z + b_0 \end{aligned}$$

whence $q(z) = Q$.

Hence Newton's method for finding a *real* zero of a polynomial is

$$\begin{aligned} x_{n+1} &= x_n - p(x_n)/p'(x_n) \\ &= x_n - P/Q \end{aligned}$$

Within a program, Q could be calculated by the function *poly* of Fig. 3.13 if the coefficients b_i were available. Consequently, when P is calculated, it is prudent to save the b_i produced. Computation of P is therefore performed by a routine similar to *poly* but which retains the coefficients of the quotient polynomial. In the procedure of Fig. 3.14, which applies Newton's method to a polynomial, this routine is called *EvaluatePoly*. Because it is returning more than one value of interest it is now a procedure rather than a function.

The procedure of Fig. 3.14 is modelled on the earlier Newton procedure of Fig. 3.12. The coefficients of the polynomial are supplied as elements 0 through n of a vector a and the coefficients of the quotient polynomial are returned as elements 0 through $n-1$ of a variable vector parameter *newa*. The function parameters f and *fdashed* are not required. The two procedure bodies are the same except that $f(x)$ and $f'(x)$ are calculated differently and, within *PolyNewton*, the coefficients of the reduced polynomial are assigned to *newa*.

```
procedure PolyNewton
  (a : vectors;   n : subs;
   x0, xtol : real;   maxits : posints;
   var root, fatroot : real;   var noofits : posints;
   var newa : vectors;   var outcome : newtoutcomes);
const
   assumedzero = 1E-20;
var
   state : newtstates;
   x, oldx, fx, fdx : real;
   itcount : 0. .maxint;
   b : vectors;

  procedure EvaluatePoly
    (a : vectors;   n : subs;   x : real;
     var p : real;   var b : vectors);
  var
     i : subs;
  begin
    b[n - 1] := a[n];
    for i := n-2 downto 0 do
      b[i] := b[i+1] * x + a[i + 1];
    p := b[0] * x + a[0]
  end { Evaluate Poly };
```

```
function poly (a : vectors;   n : subs;   x : real) : real;
  var
    i : subs;
    p : real;
begin
  p := a[n];
  for i := n-1 downto 0 do
    p := p * x + a[i];
  poly := p
end { poly };

begin { Poly Newton }
  x := x0;   EvaluatePoly (a, n, x, fx, b);
  itcount := 0;   state := iterating;
  repeat
    if abs(x) <= assumedzero then state := toonearzero else
    begin
      itcount := itcount + 1;
      fdx := poly (b, n-1, x);
      if abs(fdx) <= assumedzero then state := tooflat else
      begin
        oldx := x;   x := x - fx/fdx;
        EvaluatePoly (a, n, x, fx, b);
        if abs( (x-oldx)/oldx) <= xtol then state := withintol else
          if itcount = maxits then state := maxitsreached
      end
    end
  until state <> iterating;
  outcome := state;   noofits := itcount;
  root := x;   fatroot := fx;
  newa := b
end { Poly Newton };
```

Figure 3.14

If z is a zero of $p(x)$ then the remainder in (3.14) is zero and so

$$p(x) = (x-z)q(x)$$

and the remaining zeros of $p(x)$ are the zeros of $q(x)$. This means that $q(x)$ can be used to determine the other zeros of $p(x)$. This process is known as deflation. When a zero of $q(x)$ has been determined, a further deflation can take place and the process repeated until only a linear or quadratic factor remains. This must be done with care, however, since any zero of $p(x)$, found using Newton's method, can only be approximate. Hence the coefficients of

```
const
   tol = 1E−6;
   itbound = 50;
   n = 25;

type
   posints = 1 . . maxint;
   subs = 0 . . n;
   vectors = array [subs] of real;
   newtstates = (iterating, withintol,
                  maxitsreached, toonearzero, tooflat);
   newtoutcomes = withintol . .  tooflat;


var
   its : posints;
   outcome : newtoutcomes;
   i, m : subs;
   a, b, dummyvec : vectors;
   x, fx, dummyx, dummyf : real;
   . . .

{ Find smallest zero of polynomial a (of degree m)
  and produce deflated polynomial b (of degree m−1) }
PolyNewton (a, m, 0, tol, itbound, x, fx, its, b, outcome);
writeln ('root  ', x :10:6,
         '    f at root  ', fx :10, '    iterations  ', its :3);

for i := m − 1 downto 3 do
begin
   { Find smallest zero, x', of polynomial b (of degree i) }
   PolyNewton (b, i, 0, tol, itbound, x, fx, its, dummyvec, outcome);

   { Refine x' to x" by supplying x' as initial guess for a zero of a }
   PolyNewton (a, m, x, tol, itbound, x, fx, its, dummyvec, outcome);
   writeln ('root  ', x :10:6,
            '    f at root  ', fx :10, '    iterations  ', its :3);

   { Refine b by supplying x " as the zero }
   PolyNewton (b, i, x, tol, itbound, dummyx, dummyf, its, b, outcome)
end
```

Figure 3.15

$q(x)$ are slightly in error and any computed zero of $q(x)$ even more so. Therefore, in order to avoid a massive accumulation of numerical errors, any zero found from a deflated polynomial *must* be substituted into $p(x)$ and refined using a few steps of Newton's method. The errors inherent in this process can be minimized by trying to find the zeros in order of increasing magnitude, starting with the smallest.

The program fragment of Fig. 3.15 illustrates the process of finding successive zeros of a polynomial of degree m until the deflated polynomial is quadratic. As a simplification, no notice is taken of *outcome*; in practice the process should be terminated if any step fails to converge.

The coefficients of the polynomial p, of degree m, are stored in an array a and we shall talk of the array as though it were the polynomial itself. First the smallest zero of a is computed by applying the Newton process with the initial approximation 0. This gives the zero x and a polynomial b of degree $m - 1$, whose zeros are the remaining zeros of a. A loop then repeatedly finds the smallest zero of b and deflates b until only a quadratic polynomial remains. This could be done by a loop of the form

```
for i := m - 1 downto 3 do
begin
  PolyNewton (b, i, 0, tol, itbound, x, fx, its, b, outcome);
  writeln ('root ', x :10:6, . . . )
end
```

but this would give the accumulation of errors mentioned earlier.

To reduce this accumulation each zero (say x') computed from a deflated polynomial, b, is substituted as an initial approximation for the Newton process applied to the original polynomial, a. This produces a refined value, say x'', which is accepted as a zero of a. The coefficients of the reduced polynomial, b, are refined by applying the Newton method again but with x'' as the initial approximation.

3.3.2 Complex zeros

It cannot be emphasized too much that trying to find zeros of polynomials is an extremely difficult process; a small change to one of the coefficients in $p(x)$ can alter the zeros of the polynomial completely. See Wilkinson (1959) for an example of this. As can be imagined, finding complex zeros of polynomials is more involved. The procedure *PolyNewton* can be modified to use complex arithmetic and, using a complex initial guess, find a complex root of the polynomial. This approach has the disadvantage that complex arithmetic on computers is usually much slower than real arithmetic. Recall, however, that complex roots of polynomials with real coefficients occur in conjugate pairs and this suggests that we should seek a quadratic, rather than linear, factor of the polynomial. This is the basis for the method known as Bairstow's method.

The polynomial $p(x)$ of (3.12) can be written in the form

$$p(x) = (x^2-ux-v)q(x) + r(x-u) + s \tag{3.17}$$

where u and v are real constants, $q(x)$ has real coefficients and is of degree $n-2$

$$q(x) = b_{n-2}x^{n-2} + b_{n-3}x^{n-3} + \cdots + b_1x + b_0 \tag{3.18}$$

and the term

$$r(x-u) + s$$

is a linear remainder with real coefficients r and s and which has been written in this form to simplify later formulae. Substituting (3.18) into (3.17) and comparing coefficients with (3.12) we get

$$\begin{aligned}
a_n &= b_{n-2}\\
a_{n-1} &= b_{n-3} - ub_{n-2}\\
a_{n-2} &= b_{n-4} - ub_{n-3} - vb_{n-2}\\
&\ \ \vdots\\
a_2 &= b_0 - ub_1 - vb_2\\
a_1 &= r - ub_0 - vb_1\\
a_0 &= s - ur - vb_0
\end{aligned} \tag{3.19}$$

This can be rewritten as

$$\begin{aligned}
b_{n-2} &= a_n\\
b_{n-3} &= a_{n-1} + ub_{n-2}\\
b_{n-4} &= a_{n-2} + ub_{n-3} + vb_{n-2}\\
&\ \ \vdots\\
b_0 &= a_2 + ub_1 + vb_2\\
r &= a_1 + ub_0 + vb_1\\
s &= a_0 + ur + vb_0
\end{aligned} \tag{3.20}$$

For x^2-ux-v to be a factor of $p(x)$ we require $r = s = 0$ but r and s are both functions of u and v. Thus we require that the equations

$$\begin{aligned}
r(u,v) &= 0\\
s(u,v) &= 0
\end{aligned}$$

be solved for the unknown values u and v.

As yet, we do not know how to solve these equations but, in Section 3.4, a method based on Newton's method is described for solving precisely this type of problem. A correction $\boldsymbol{\delta w}_j$ to the jth iterate

$$\mathbf{w}_j = \begin{bmatrix} u_j \\ v_j \end{bmatrix}$$

can be found by solving the system of *linear* equations

$$J_j \boldsymbol{\delta w}_j = \begin{bmatrix} r(u_j,v_j) \\ s(u_j,v_j) \end{bmatrix} \tag{3.21}$$

where

$$J_j = \begin{bmatrix} (r)_u & (r)_v \\ (s)_u & (s)_v \end{bmatrix}$$

in the notation of Section 3.4 where, for example, $(r)_u$ is the first partial derivative of $r(u,v)$ with respect to u.

It would appear that we cannot evaluate these partial derivatives since we have no explicit representation for r and s in terms of u and v but merely the recurrence (3.20). However, the recurrence relation itself can be differentiated. For example, if the first few equations of (3.20) are differentiated with respect to u then, since b_{n-2} is a constant,

$$(b_{n-2})_u = 0$$

and the recurrence relationship

$$\begin{aligned}
(b_{n-3})_u &= b_{n-2} \\
(b_{n-4})_u &= b_{n-3} + u(b_{n-3})_u \\
(b_{n-5})_u &= b_{n-4} + u(b_{n-4})_u + v(b_{n-3})_u \\
&\;\;\vdots \\
(b_0)_u &= b_1 + u(b_1)_u + v(b_2)_u \\
(r)_u &= b_0 + u(b_0)_u + v(b_1)_u \\
(s)_u &= r + u(r)_u + v(b_0)_u
\end{aligned}$$

is obtained. Defining

$$c_k = (b_{k-2})_u$$

with

$$c_1 = (r)_u \text{ and } c_0 = (s)_u$$

the relationship can be rewritten as

$$\begin{aligned}
c_{n-1} &= b_{n-2} \\
c_{n-2} &= b_{n-3} + uc_{n-1} \\
c_{n-3} &= b_{n-4} + uc_{n-2} + vc_{n-1} \\
&\vdots \\
c_1 &= b_0 + uc_2 + vc_3 \\
c_0 &= r + uc_1 + vc_2
\end{aligned} \tag{3.22}$$

so that J_j now has the form

$$J_j = \begin{bmatrix} c_1 & (r)_v \\ c_0 & (s)_v \end{bmatrix}$$

By differentiating (3.20) with respect to v and setting

$$d_k = (b_{k-3})_v$$

the following relationships can be derived:

$$\begin{aligned}
d_{n-1} &= b_{n-2} \\
d_{n-2} &= b_{n-3} + ud_{n-1} \\
d_{n-3} &= b_{n-4} + ud_{n-2} + vd_{n-1} \\
&\vdots \\
d_2 &= b_1 + ud_3 + vd_4 \\
d_1 &= b_0 + ud_2 + vd_3
\end{aligned} \tag{3.23}$$

so that

$$J_j = \begin{bmatrix} c_1 & d_2 \\ c_0 & d_1 \end{bmatrix}$$

Note, however, that the relations (3.22) and (3.23) produce the same values, i.e.

$$d_k = c_k \qquad \text{for } k = n-1, n-2, \ldots, 1$$

and so (3.23) is redundant and J_j can be written as

$$J_j = \begin{bmatrix} c_1 & c_2 \\ c_0 & c_1 \end{bmatrix}$$

Now we can write down the Bairstow algorithm for determining a quadratic factor of a polynomial, $p(x)$.

(i) Obtain initial estimates, u_0 and v_0, of the values u and v.
(ii) Calculate r and s using (3.20) and the derivatives c_2, c_1 and c_0 using (3.22).
(iii) Calculate the correction $\boldsymbol{\delta w}_j$ using (3.21) and (3.22) to give

$$\delta u_j = -\frac{rc_1 - sc_2}{c_1^2 - c_0c_2}$$

$$\delta v_j = -\frac{sc_1 - rc_0}{c_1^2 - c_0c_2}$$

(iv) If each element of $\boldsymbol{\delta w}_j$ satisfies an appropriate relative error test, then stop with

$$\mathbf{w} = \begin{bmatrix} u \\ v \end{bmatrix} = \mathbf{w}_j + \boldsymbol{\delta w}_j$$

and find the zeros of the resulting quadratic, $x^2 - ux - v$; otherwise set

$$\mathbf{w}_{j+1} = \mathbf{w}_j + \boldsymbol{\delta w}_j$$

and return to (ii) with j replaced by $j+1$.

The procedure of Fig. 3.16 implements the method. The two types *bairstates* and *bairoutcomes* are assumed to be

```
bairstates = (iterating, withintol,
              maxitsreached, Jnearzero, toonearzero);
bairoutcomes = withintol . . toonearzero;
```

```
procedure Bairstow
  (a : vectors;
   u, v : real;   n : subs;   tol : real;   maxits : posints;
   var newu, newv : real;   var noofits : posints;
   var b : vectors;   var outcome : bairoutcomes);
  const
    assumedzero = 1E–20;
  var
    r, s, b0, b1, c0, c1, c2, J, du, dv : real;
    i : subs;
    itcount : 0 . . maxint;
    state : bairstates;
```

```
begin
  itcount := 0;   state := iterating;
  repeat
    itcount := itcount + 1;
    b0 := a[n];   r := a[n-1] + u * b0;
    c1 := b0;   c0 := r + u * c1;
    for i := n-2 downto 1 do
    begin
      b1 := b0;   b0 := r;
      r := a[i] + u * b0 + v * b1;
      c2 := c1;   c1 := c0;
      c0 := r + u * c1 + v * c2
    end;
    s := a[0] + u * r + v * b0;
    J :- c0 * c2 - sqr(c1);
    if abs(J) <= assumedzero then state := Jnearzero else
    begin
      du := (r * c1 - s * c2) / J;
      dv := (s * c1 - r * c0) / J;
      u := u + du;   v := v + dv;
        state := toonearzero else
        if (abs(du/u)<=tol) and (abs(dv/v)<=tol) then
            state := withintol else
            if itcount = maxits then state := maxitsreached
      if itcount = maxits then state := maxitsreached
    end
  until state <> iterating;

  outcome := state;
  noofits := itcount;   newu := u;   newv := v;
  b[n - 2] := a[n];
  b[n - 3] := a[n - 1] + u * b[n - 2];
  for i := n - 4 downto 0 do
    b[i] := a[i] + u * b[i+1] + v * b[i+2]
end { Bairstow };
```

Figure 3.16

Having found a quadratic factor, the process of deflation is as for a linear factor with the coefficients of the reduced polynomial calculated as part of the algorithm. Again, each factor must be verified.

3.4 Systems of non-linear equations

The problem of solving large systems of non-linear equations is extremely complicated and beyond the scope of this book. However, the methods involved in solving such problems are, essentially, variants of the technique, described here, for solving two non-linear equations in two unknowns. The method, Newton's method applied to the system of equations

$$f(x,y) = 0, \qquad g(x,y) = 0 \tag{3.24a}$$

is an iteration designed to find the values α and β such that

$$f(\alpha,\beta) = 0, \qquad g(\alpha,\beta) = 0 \tag{3.24b}$$

simultaneously.

Assuming that (x_n,y_n) is an iterate which is sufficiently close to the root (α,β), then (3.24b) can be expanded about (x_n,y_n), using two-dimensional Taylor's series, to give

$$\begin{aligned} 0 &= f(\alpha,\beta) = f + (\alpha - x_n)f_x + (\beta - y_n)f_y + \cdots \\ 0 &= g(\alpha,\beta) = g + (\alpha - x_n)g_x + (\beta - y_n)g_y + \cdots \end{aligned} \tag{3.25}$$

where, for example, f is $f(x_n,y_n)$ and f_x is the partial derivative of $f(x,y)$ with respect to x evaluated at (x_n,y_n).

The assumption that (x_n,y_n) is sufficiently close to (α,β) implies that terms in $(\alpha - x_n)^2$ and $(\beta - y_n)^2$ and higher orders can be neglected. If we define the vector $\boldsymbol{\delta x}_n$ by

$$\boldsymbol{\delta x}_n = \begin{bmatrix} \alpha - x_n \\ \beta - y_n \end{bmatrix}$$

then (3.25) can be rewritten in the form

$$J_n \boldsymbol{\delta x}_n = -\boldsymbol{b}_n \tag{3.26}$$

where

$$J_n = \begin{bmatrix} f_x & f_y \\ g_x & g_y \end{bmatrix}_{(x_n,y_n)}$$

and

$$\boldsymbol{b}_n = \begin{bmatrix} f \\ g \end{bmatrix}_{(x_n,y_n)}$$

J_n is known as the *Jacobian* matrix.

As a consequence of having neglected terms in the series expansion, the solution of (3.26) is not exact. However, by redefining the vector $\boldsymbol{\delta x}_n$ as

$$\boldsymbol{\delta x}_n = \begin{bmatrix} x_{n+1} - x_n \\ y_{n+1} - y_n \end{bmatrix} \tag{3.27}$$

(3.25) can be used as an iterative formula to obtain new estimates, x_{n+1} and y_{n+1}, of the roots. Formula (3.26) can be viewed as an iteration and is the two-dimensional analogue of Newton's method. Under favourable circumstances the iteration defined by (3.26) and (3.27) gives a sequence of iterates which converges to (α,β).

For two equations in two unknowns the solution of (3.26) is particularly simple. A straightforward elimination shows that

$$\begin{aligned} x_{n+1} - x_n &= -\frac{fg_y - gf_y}{f_xg_y - g_xf_y} \\ y_{n+1} - y_n &= -\frac{gf_x - fg_x}{f_xg_y - g_xf_y} \end{aligned} \tag{3.28}$$

An improvement in the rate of convergence may be obtained by evaluating the second equation in (3.28) using the latest value, x_{n+1}, calculated in the first equation. This necessitates recalculating each term on the right-hand side of the equation. The procedure of Fig. 3.17 adopts this approach. The function values at the computed root are not returned. The data types *simstates* and *simoutcomes* are exactly *bairstates* and *bairoutcomes* of Section 3.3.2.

```
procedure TwoSimEqns
  (function f (x, y :real) : real;
   function fx (x, y : real) : real;
   function fy (x, y : real) : real;
   function g (x, y : real) : real;
   function gx (x, y : real) : real;
   function gy (x, y : real) : real;
   x0, y0, tol : real;   maxits : posints;
   var xroot, yroot : real;
   var noofits : posints;   var outcome : simoutcomes);
  const
    assumedzero = 1E−20;
  var
    x, oldx, y, oldy, J : real;
    itcount : 0 . . maxint;
    state : simstates;
```

```
begin
  x := x0;   y := y0;
  itcount := 0;   state := iterating;
  repeat
    if (abs(x)<=assumedzero) or (abs(y)<=assumedzero) then
      state := toonearzero else
    begin
      J := fx(x,y) * gy(x,y) - gx(x,y) * fy(x,y);
      if abs(J)<=assumedzero then state := Jnearzero else
      begin
        itcount := itcount + 1;   oldx := x;   oldy := y;
        x := x - (f(x,y) * gy(x,y) - g(x,y) * fy(x,y)) / J;
        J := fx(x,y) * gy(x,y) - gx(x,y) * fy(x,y);
        if abs(J)<=assumedzero then state := Jnearzero else
        begin
          y := y - (g(x,y) * fx(x,y) - f(x,y) * gx(x,y)) / J;
          if (abs((x-oldx)/oldx) <= tol) and
            (abs((y-oldy)/oldy) <= tol) then state := withintol else
            if itcount = maxits then state := maxitsreached
        end
      end
    end
  until state <> iterating;
  outcome := state;
  noofits := itcount;   xroot := x;   yroot := y
end { Two sim eqns };
```

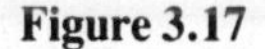

Figure 3.17

Again it can be shown, as for a single equation, that the order of the Newton iteration (3.28) is 2. However, it must be emphasized that solving a system of non-linear equations is very much more difficult than solving a single equation. For single equations good information about starting values can always be obtained by using, for example, the bisection method, and so interest usually centres around the order of convergence. For a system of equations, however, this information is not readily available and, in general, the main point of interest is whether or not the method will converge at all. For this reason it is essential that a careful check be kept on the progress of the iteration to discover any lack of convergence as soon as possible. For example, during a trial run, the maximum number of iterations permitted could be kept small and successive iterates printed.

3.5 Exercises

1 Write a procedure to perform the regula falsi iteration. Examine the performance of the method when applied to finding a root of the function

$$f(x) = e^{-x} \ln(x)$$

in the interval [0.5, 4]. Print out clearly the result of each iteration.

2 Use the bisection, secant and Newton methods to determine the root of the function in Exercise 1. Should any iterations not converge, then experiment by taking different initial intervals and starting values and try to explain the behaviour of each method.

3 Compare the performance of the bisection, secant, regula falsi and Newton methods on the function

$$f(x) = x^3 - x + 1$$

of Fig. 3.11 using a variety of starting intervals and values.

4 Apply Aitken's Δ^2 process to the sequences of iterates produced in Exercises 1 and 2.

5 Prove Theorem 3.3 using the same techniques as for Theorem 3.2.

6 Derive several fixed-point iterations based on the function

$$f(x) = x^2 + 2x \sin(x) - 1$$

which has a root in the interval [0,1]. Check whether or not they are convergent by applying the result of Theorem 3.2, and verify your conclusions by programming the methods and running the programs.

7 Write and run a program to use the fragment of Fig. 3.15 to determine all the roots of the polynomial

$$p(x) = (x-6)(x-5)(x-4)(x-3)(x-2)(x-1)$$

8 Modify the program of Exercise 7 so that each root is *not* refined and compare the results with those of the program in the previous example.

9 Write a program to use Bairstow's method and deflation to find all the roots of a polynomial and run it for

$$p(x) = x^5 + 14x^2 + 12x + 12$$

Chapter 4 LINEAR ALGEBRAIC EQUATIONS

This chapter is concerned with the problem of determining the values $x_1, x_2, \ldots, x_n$ which satisfy the following system of n linear equations

$$
\begin{aligned}
a_{11}x_1 + a_{12}x_2 + \cdots + a_{1n}x_n &= b_1 \\
a_{21}x_1 + a_{22}x_2 + \cdots + a_{2n}x_n &= b_2 \\
&\vdots \\
a_{n1}x_1 + a_{n2}x_2 + \cdots + a_{nn}x_n &= b_n
\end{aligned}
\tag{4.1}
$$

Equation (4.1) can be written in the form

$$A\boldsymbol{x} = \boldsymbol{b} \tag{4.2}$$

where A is an $n \times n$ matrix whose (i,j)th element, a_{ij}, is the jth coefficient in the ith equation of (4.1). We have already seen an example of this type of problem in Chapter 3 when it was necessary to solve the two equations of (3.24a) simultaneously. In this case n had the value 2 and writing down the solution was simple.

It is assumed, throughout this chapter, that the reader is familiar with basic linear algebra including the result that the solution to equation (4.2) is unique provided the matrix A is non-singular, i.e. that the determinant of A is non-zero. It will normally be assumed that this is the case but various tests will be included to detect possible singularity. We refer the reader to McKeown and Rayward-Smith (1982) as a general background text.

This chapter splits into two sections. The first discusses some *direct* methods, these reduce the given equations to a form from which the solution can be obtained simply. The second presents several *iterative* methods which bear some similarity to the methods discussed in Chapter 3; an approximation is made and successively refined until it is acceptably close to the solution.

4.1 Direct methods

If the matrix in equation (4.2) is diagonal ($a_{ij} = 0$ for $i \neq j$) then the solution can be written down immediately:

$$x_i = \frac{b_i}{a_{ii}}, \qquad a_{ii} \neq 0 \quad \text{for } i = 1,2,\ldots,n$$

The solution can be obtained almost as easily if A is triangular and no diagonal element, a_{ii}, is zero. For example, if the system of equations is upper triangular, in the form

$$\begin{aligned}
a_{11}x_1 + a_{12}x_2 + \quad \cdots \quad + a_{1n}x_n &= b_1 \\
a_{22}x_2 + \quad \cdots \quad + a_{2n}x_n &= b_2 \\
&\;\;\vdots \\
a_{n-1n-1}x_{n-1} + a_{n-1n}x_n &= b_{n-1} \\
a_{nn}x_n &= b_n
\end{aligned} \tag{4.3}$$

then x_n, the nth element of the solution vector, is immediately determined from the nth equation of (4.3), viz.

$$x_n = \frac{b_n}{a_{nn}}$$

Once x_n has been determined, x_{n-1} is the only unknown in the $(n-1)$st equation and can be obtained directly by rewriting this equation in the form

$$x_{n-1} = \frac{b_{n-1} - a_{n-1n}x_n}{a_{n-1n-1}}$$

In general,

$$x_i = \frac{b_i - a_{ii+1}x_{i+1} - \cdots - a_{in}x_n}{a_{ii}} \qquad \text{for } i = n-1, n-2, \ldots, 1 \tag{4.4}$$

Thus all the x_i $\{i = n, n-1, \ldots, 1\}$ can be determined by this *back-substitution* process.

If, within a Pascal program, x and b are one-dimensional arrays and the coefficient matrix is represented by a two-dimensional array a, the following program fragment will perform the back-substitution process.

```
x[n] := b[n] / a[n,n];
for i := n-1 downto 1 do
begin
  top := b[i];
  for j := i+1 to n do
    top := top - a[i,j] * x[j];
  x[i] := top / a[i,i]
end
```

4.1.1 Gaussian elimination

The method of Gaussian elimination comprises two stages. A process of *elimination* systematically reduces a non-singular system of equations of the form (4.1) to the upper triangular form (4.3). The solution is then produced by the back-substitution process described in (4.4).

The elimination procedure involves the subtraction of a multiple of one equation from another. The proof of the following theorem is left as an exercise for the reader.

Theorem 4.1

Let $\boldsymbol{Ax}=\boldsymbol{b}$ be a given linear system and suppose that the system is subjected to a sequence of operations of the following kind:

(i) multiplication of one equation by a non-zero constant,
(ii) addition (subtraction) of a multiple of one equation to (from) another,
(iii) interchange of two equations.

If this sequence of operations produces the new system $\boldsymbol{A}^*\boldsymbol{x}=\boldsymbol{b}^*$, then the system $\boldsymbol{Ax}=\boldsymbol{b}$ and $\boldsymbol{A}^*\boldsymbol{x}=\boldsymbol{b}^*$ are equivalent and, in particular, have the same solution.

The method will be described by considering an example. Suppose that the solution to the three equations

$$\begin{aligned} 2x_1 + x_2 + 3x_3 &= 11 \\ 4x_1 + 3x_2 + 10x_3 &= 28 \\ 2x_1 + 4x_2 + 17x_3 &= 31 \end{aligned} \qquad (4.5)$$

is required. The first step in the 'triangularisation' of this system of equations is to eliminate x_1 from the second and third equations of (4.5). This can be achieved by subtracting suitable multiples of the first equation from the other two. During this process, the first equation is called the *pivotal* equation and its first coefficient (2) is called the *pivot*. To eliminate x_1 from the second equation, the *multiplier* must be 2 and to eliminate x_1 from the third equation, the multiplier must be 1. This elimination gives the reduced set of equations

$$\begin{array}{lrl} & 2x_1 + x_2 + 3x_3 &= 11 \\ 2 & x_2 + 4x_3 &= 6 \\ 1 & 3x_2 + 14x_3 &= 20 \end{array}$$

The two multipliers have been recorded, to the left of the equations. The system is reduced to triangular form if x_2 is now eliminated from the third equation. The second equation now becomes the pivotal equation and its diagonal element (1), the coefficient of x_2, becomes the pivot. The multiplier

3/1 is formed and this multiple of the second equation is subtracted from the third to give

$$\begin{array}{lr} & 2x_1 + x_2 + 3x_3 = 11 \\ 2 & x_2 + 4x_3 = \ 6 \\ 1 \quad 3 & 2x_3 = \ 2 \end{array} \tag{4.6}$$

and, again, the multiplier has been recorded. These equations can now be solved using back-substitution to give the solution

$$\boldsymbol{x} = \begin{bmatrix} 3 \\ 2 \\ 1 \end{bmatrix}$$

To see how the method can be written down for a general matrix, A, let us assume that the process is partially completed and we have reached the kth stage where the system of equations has the form

$$\begin{array}{llll} a_{11}x_1 + a_{12}x_2 + & \cdots & + a_{1n}x_n & = b_1 \\ \quad a_{22}x_2 + & \cdots & + a_{2n}x_n & = b_2 \\ & & & \ \ \vdots \\ \quad\quad a_{kk}x_k \quad + & \cdots & + a_{kn}x_n & = b_k \\ \quad\quad a_{k+1k}x_k + & \cdots & + a_{k+1n}x_n & = b_{k+1} \\ & & & \ \ \vdots \\ \quad\quad a_{nk}x_k + & \cdots & + a_{nn}x_n & = b_n \end{array} \tag{4.7}$$

The elimination has been carried out on the first $k-1$ columns of A and now it is the turn of the kth column. It should be noted that only the coefficients in the first equation correspond to those of the original system, all the others have been modified during the elimination. The process is the same for each of the rows, $i = k+1,\ldots,n$; firstly the multiplier, a_{ik}/a_{kk}, is formed and then that multiple of row k is subtracted from row i so that the following steps are taken for each value $i = k+1,\ldots,n$

(i) form the multiplier

$$m_{ik} = \frac{a_{ik}}{a_{kk}}$$

(ii) for each value of $j = k+1,\ldots,n$ form the new value

$$a_{ij} = a_{ij} - m_{ik} * a_{kj}$$

(iii) finally form the new value

$$b_i = b_i - m_{ik} * b_k$$

This process is applied for $k = 1,2,...,n-1$. Burden *et al.* (1981, p.271) perform an operation count and show that the total number of multiplications and divisions used in the elimination is

$$\frac{n^3 + 3n^2 + n}{3}$$

This shows that the amount of computation grows rapidly as the number of equations increases.

A procedure implementing Gaussian elimination is given in Fig. 4.1. The data types *nbyn*, *vectors* and *subs* were defined in Section 1.3.3 (p.13) and *n* is assumed to be a global constant. The coefficient matrix A is represented by a two-dimensional array *a* and each access to an individual element of A is achieved by a doubly subscripted expression. The updated values of *a* and *b* are of no interest outside the procedure and so these are *value* parameters. The solution vector will be returned via *x* and so this must be a *variable* parameter.

```
procedure Gauss (a : nbyn;   b : vectors;   var x : vectors);

    { assumes no pivot is ever zero }

  var
    i, j, k : subs;
    pivot, mult, top : real;

begin
  for j := 1 to n - 1 do
  begin
      { take row j as the pivotal row and
        eliminate column j from rows j + 1 onwards }
    pivot := a[j,j];
    for i := j + 1 to n do
    begin
      mult := a[i,j] / pivot;
        { now subtract mult * pivotal row from row i }
      for k := j+1 to n do
        a[i,k] := a[i,k] - mult * a[j,k];
      b[i] := b[i] - mult * b[j]
    end
  end;
```

```
    { now perform back substitution }
  x[n] := b[n] / a[n,n];
  for i := n-1 downto 1 do
  begin
    top := b[i];
    for k := i+1 to n do
      top := top - a[i,k] * x[k];
    x[i] := top / a[i,i]
  end
end { Gauss };
```

Figure 4.1

Chapter 1 presented some slicing techniques to avoid double-subscripting and, hence, improve program efficiency. One involved passing rows of the structure as parameters to procedures. This is appropriate in the present context because all operations on A are applied row-wise during both the elimination and the back-substitution.

The procedure of Fig. 4.2 performs Gaussian elimination by slicing the coefficient matrix. The procedure *SubtractRow* is given two rows, the multiplier and the position in the rows at which the subtraction is to commence and it performs the appropriate subtraction, now using single subscripting. The parameter u returns the updated row and so must be transferred as a *variable*. The function *dotprod* takes two vectors and a subscript k and computes the dot product of that portion of the two vectors comprising elements $k, k+1, \ldots, n$.

```
procedure SlicedGauss (a : nbyn;   b : vectors;
                         var x : vectors;   var success : boolean);

    { checks for a zero pivot }

  label 999;

  const
    assumedzero = 1E-20;

  var
    i, j : subs;
    pivot, mult : real;

  function dotprod (u, v : vectors;   k : subs) : real;
      { u[k. .n] . v[k. .n] }
    var
      i : subs;
      sum : real;
```

```
  begin
    sum := 0;
    for i := k to n do
      sum := sum + u[i] * v[i];
    dotprod := sum
  end { dot prod };

  procedure SubtractRow
      (var u : vectors;   v : vectors;   m : real;   k : subs);
        { u[k. .n] := u[k. .n] - m * v[k. .n] }
    var
      i : subs;
  begin
    for i := k to n do
      u[i] := u[i] - m * v[i]
  end { Subtract Row };

begin { Sliced Gauss }
  for j := 1 to n - 1 do
  begin
    pivot := a[j, j];
    if abs (pivot) <= assumedzero then
    begin
      success := false;   goto 999
    end;
    for i := j + 1 to n do
    begin
      mult := a[i,j] / pivot;
      if abs (mult) > assumedzero then
      begin
        SubtractRow (a[i], a[j], mult, j + 1);
        b[i] := b[i] - mult * b[j]
      end
    end
  end;
  success := abs(a[n,n]) > assumedzero;
  if success then
  begin
    x[n] := b[n] / a[n,n];
    for i := n - 1 downto 1 do
      x[i] := (b[i] - dotprod (a[i], x, i + 1)) / a[i,i]
  end;
999 :
end { Sliced Gauss };
```

Figure 4.2

The procedure of Fig. 4.2 checks for a zero pivot. If the absolute value of a pivot is less than a small value *assumedzero*, a boolean variable parameter *success* is set to *false* and control leaves the procedure immediately. This is achieved with a goto-statement. It was stressed in Section 1.1.11 that the use of such a statement can be dangerous and is not to be encouraged. One context in which its use is perhaps justified is to exit, under exceptional circumstances, from a loop or set of nested loops which are best written as for-statements. Such is the present context: we are testing for a condition which we hope will never occur but which, if it does occur, must cause exit from a loop which we would like to write as a for-statement.

The procedure can be written without the goto-statement and this has been done in Fig. 4.3. Notice that the outermost for-statement has been replaced by a repeat-statement and consequently the fact that j is normally expected to take successive values from 1 to $n-1$ is no longer immediately apparent. It could be argued that this approach is therefore less transparent than that of Fig. 4.2 and so, for the remainder of this chapter, we shall adopt the approach of Fig. 4.2.

```
procedure SlicedGauss (a : nbyn;   b : vectors;
                          var x : vectors;   var success : boolean);

    { checks for a zero pivot }

  const
    assumedzero = 1E-20;

  var
    i, j : subs;
    pivot, mult : real;
    state : (incrementingj, jisnminus1, zeropivot);

  function dotprod (u, v : vectors;   k : subs) : real;
      { u[k. .n] . v[k. .n] }
    var
      i : subs;
      sum : real;
  begin
    sum := 0;
    for i := k to n do
      sum := sum + u[i] * v[i];
    dotprod := sum
  end { dot prod };
```

```
procedure SubtractRow
    (var u : vectors;   v : vectors;   m : real;   k : subs);
      { u[k. .n] := u[k. .n] - m * v[k. .n] }
  var
    i : subs;
begin
  for i := k to n do
    u[i] := u[i] - m * v[i]
end { Subtract Row };

begin { Sliced Gauss }
  j := 1;   state := incrementingj;
  repeat
    pivot := a[j,j];
    if abs(pivot) <= assumedzero then state := zeropivot else
    begin
      for i := j + 1 to n do
      begin
        mult := a[i,j] / pivot;
        if abs(mult) > assumedzero then
        begin
          SubtractRow (a[i], a[j], mult, j + 1);
          b[i] := b[i] - mult * b[j]
        end
      end;
      if j = n - 1 then state := jisnminus1 else
        j := j + 1
    end
  until state <> incrementingj;

  case state of
    zeropivot : success := false;
    jisnminus1 :
      begin
        success := abs(a[n,n]) > assumedzero;
        if success then
        begin
          x[n]:=b[n]/a[n,n]
          for i := n - 1 downto 1 do
            x[i] := (b[i] - dotprod (a[i], x, i + 1)) / a[i,i]
        end
      end
  end { case }
end { Sliced Gauss };
```

Figure 4.3

The value *a*[*n*,*n*] could become zero without any of the pivots failing the test and so, before the back-substitution, it is compared with *assumedzero*. In addition, row subtraction for small multipliers is avoided; there is little point in performing the row subtraction if the multiplier is effectively zero.

Another technique to avoid double-subscripting, mentioned in Chapter 1, is the use of records. It was stressed that an algorithm should be defined before the data structure representations are chosen. The data structures are then chosen to facilitate the algorithm. In the context of Gaussian elimination we see that elements of A are processed row-wise and an element of $\boldsymbol{b}$ is processed along with each row of A. This suggests that the best representation is a vector of records where each record includes a row of A and one element of $\boldsymbol{b}$. The solution vector $\boldsymbol{x}$ is referred to only in the back-substitution process and then several elements of $\boldsymbol{x}$ are used in conjunction with one row of A. This suggests that $\boldsymbol{x}$ should be stored separately. The data types used in addition to *subs* and *vectors* of Section 1.3.3, (p.13) are as follows:

```
rows = record
          coeff : vectors;
              b : real
       end { rows };

rowvectors = array [subs] of rows;
```

The procedure of Fig. 4.4 performs Gaussian elimination assuming these data types. It includes the testing of pivots and multipliers introduced in Fig. 4.2 and performs a further action. Each multiplier m_{ij} overwrites the

---

```
procedure RecordGauss (var eqn : rowvectors;
                       var x : vectors;   var success : boolean);

  label 999;

  const
    assumedzero = 1E-20;

  var
    i, j, k : subs;
    pivot, mult, top : real;

  procedure Subtractrow
      (var u : rows;   v : rows;   m : real;   k : subs);
        { u.coeff[k. .n] := u.coeff[k. .n] - m * v.coeff[k. .n]
                     u.b := u.b - m * v.b }
    var
      i : subs;
```

```
    begin
      with u do
      begin
        for i := k to n do
          coeff[i] := coeff[i] - m * v.coeff[i];
        b := b - m * v.b
      end
    end { Subtract Row };

  begin {Record Gauss }
    for j := 1 to n - 1 do
    begin
      pivot := eqn[j].coeff[j];
      if abs (pivot) <= assumedzero then
      begin
        success := false;    goto 999
      end;
      for i := j + 1 to n do
      with eqn[i] do
      begin
        mult := coeff[j] / pivot;
        if abs (mult) > assumedzero then
        begin
          coeff[j] := mult;
          SubtractRow (eqn[i], eqn[j], mult, j + 1)
        end else
          coeff[j] := 0
      end
    end;
    success := abs(eqn[n].coeff[n]) > assumedzero;
    if success then
    begin
      with eqn[n] do
        x[n] := b / coeff[n];
      for i := n - 1 downto 1 do
      with eqn[i] do
      begin
        top := b;
        for k := i + 1 to n do
          top := top - coeff[k] * x[k];
        x[i] := top / coeff[i]
      end
    end
  end { Record Gauss };
```

Figure 4.4

coefficient a_{ij} unless the multiplier is smaller than *assumedzero*, in which case the element a_{ij} is set to zero. Overwriting these coefficients has no effect on the method because, once the multiplier m_{ij} has been computed, the element a_{ij} is never referred to again. However, as will be shown in Section 4.1.4, the lower triangular matrix of multipliers, together with the final upper triangular form of the coefficient matrix, can be useful. So that the updated information is returned, *eqn* is a *variable* parameter.

4.1.2 Pivoting strategy

The procedures of Figs. 4.3 and 4.4 abandon computation if a zero pivot is encountered. But a system of equations can have a zero on the diagonal and yet have a perfectly well-defined solution. For example, the system of equations

$$\begin{aligned} x_2 &= 1 \\ x_1 + x_2 &= 2 \end{aligned}$$

has a zero on the diagonal but the solution, $x_1 = 1$ and $x_2 = 1$, is well defined. The problem can be resolved by reordering the equations to give

$$\begin{aligned} x_1 + x_2 &= 2 \\ x_2 &= 1 \end{aligned}$$

which can be solved using back-substitution since the system is now upper-triangular. So it appears that the problem of a zero pivot can be overcome, say, by looking down the pivotal column until a non-zero element is met and then swapping this row with the existing pivotal row. This will work, but there is one further consideration. Since problems arise if a pivot is zero it seems reasonable to assume that they may also arise if the pivot is small compared to the other elements of the matrix and this is, in fact, the case. So, rather than accept the first non-zero element in the pivotal column, the whole column, below the pivot, should be scanned to determine the element of maximum modulus and then the rows exchanged, if necessary, so that this element becomes the pivot. This process is known as *partial pivoting* and is a fundamental step in the solution of systems of linear equations.

There is a process of *complete pivoting*, in which the element of greatest modulus in the submatrix, a_{ij} for $i,j \geqslant k$, in (4.7), is located, whereupon row *and* column interchanges are performed to move this element to the pivotal position. Computationally, this is extremely costly and is rarely necessary in practice; partial pivoting is usually adequate. If no non-zero element can be found in the pivotal column, then the matrix is singular and this can be used as a check for singularity.

The process of partial pivoting can be illustrated by considering the solution of equations (4.5) discussed earlier. The coefficients of x_1, in each row, are 2, 4 and 2, respectively, so that the multipliers are 2 and 1. In order to make the pivot as large as possible and thus reduce the size of the multipliers the first and second rows should be exchanged. This produces the system

$$\begin{aligned} 4x_1 + 3x_2 + 10x_3 &= 28 \\ 2x_1 + x_2 + 3x_3 &= 11 \\ 2x_1 + 4x_2 + 17x_3 &= 31 \end{aligned}$$

with 0.5 as the multiplier for both rows two and three. However, if the system of equations is very large, the process of actually swapping rows is extremely time consuming. Instead, we maintain a record of the *order* in which rows are to be considered and update the ordering as successive pivotal rows are chosen. In the ensuing example the permutation of the rows is indicated by three pointers p_1, p_2 and p_3. When we talk of row p_1 we shall mean the row currently pointed to by p_1, and similarly for rows p_2 and p_3. Initially p_1, p_2 and p_3 point to rows 1, 2 and 3 respectively.

	A			b
$p_1 \longrightarrow$	2	1	3	11
$p_2 \longrightarrow$	4	3	10	28
$p_3 \longrightarrow$	2	4	17	31

To determine the first pivot, the maximum modulus element in the first column is sought. It is located in row p_2 (row 2) so we wish to interchange rows p_1 (row 1) and p_2 (row 2). Rather than interchange the rows themselves, we merely interchange the destinations of the two pointers.

	A			b
$p_1 \searrow$ (to row 2)	2	1	3	11
$p_2 \nearrow$ (to row 1)	4	3	10	28
$p_3 \longrightarrow$	2	4	17	31

We now eliminate x_1 from all the rows 'below' the pivotal row by subtracting suitable multiples (0.5 in each case) of row p_1 (row 2) from rows p_2 (row 1) and p_3 (row 3).

	A			b
p_1	0	−0.5	−2	−3
p_2	4	3	10	28
p_3	0	2.5	12	17

Henceforth, p_1 is ignored and subsequent processing involves rows p_2 (row 1) and p_3 (row 3). The second column of these two rows is scanned to determine the pivot. This turns out to be 2.5 in row p_3 (row 3) so now rows p_2 (row 1) and p_3 (row 3) must be interchanged. Accordingly, the destinations of the two pointers are swapped.

	A			b
p_1	0	−0.5	−2	−3
p_2	4	3	10	28
p_3	0	2.5	12	17

Row p_2 (row 3) now remains unchanged and the multiplier for row p_3 (row 1) is

$$\frac{-0.5}{2.5} = -0.2$$

	A			b
p_1	0	0	0.4	0.4
p_2	4	3	10	28
p_3	0	2.5	12	17

We have now produced the upper triangular form. The first row is p_1 (row 2), the second is p_2 (row 3) and the third is p_3 (row 1).

To perform the back-substitution we work through the rows in the reverse order: p_3,p_2,p_1. From row p_3 (row 1) we have

$$x_3 = \frac{0.4}{0.4} = 1$$

From row p_2 (row 3) we have

$$x_2 = \frac{17 - 12x_3}{2.5} = \frac{5}{2.5} = 2$$

From row p_1 (row 2) we have

$$x_1 = \frac{28 - 3x_2 - 10x_3}{4} = \frac{28 - 6 - 10}{4} = \frac{12}{4} = 3$$

In practice, the multipliers would overwrite the unwanted coefficients and so the final form of the coefficient matrix would be as follows

	A			b
p_1 → row 2	0.5	−0.2	0.4	0.4
p_2 → row 3	4	3	10	28
p_3 → row 1	0.5	2.5	12	17

The procedure of Fig. 4.5 implements this process. The data structure representations are similar to those of Fig. 4.3 in that one element of $\boldsymbol{b}$ is associated with each row of coefficients but the overall form is a vector of *pointers* to rows rather than a vector of rows. The data types *rowptrs* and *ptrvecs* are defined as follows:

```
rows = record
         coeff : vectors;
             b : real
       end { rows };
rowptrs = ↑ rows;
ptrvecs = array [subs] of rowptrs;
```

```
procedure PointerGauss (var eqn : ptrvecs;
                        var x : vectors;   var success : boolean);

  const
    assumedzero = 1E-20;

  var
    i, j, k : subs;
    top, pivot, mult : real;
    pivotalrow : rowptrs;

  procedure SubtractRow
      (var u : rows;   v : rows;   m : real;   k : subs);
        { u.coeff[k. .n] := u.coeff[k. .n] - m * v.coeff[k. .n]
                     u.b := u.b - m * v.b }
    var
      i : subs;
```

```
begin
  with u do
  begin
    for i := k to n do
      coeff[i] := coeff[i] - m * v.coeff[i];
    b := b - m * v.b
  end
end { Subtract Row };

procedure SwapPivRow (var a : ptrvecs;   col : subs);
    { Scans column 'col' of rows a[col] ↑ , a[col+1] ↑ , . . . ,
      a[n] ↑  to find the pivotal row. The pointers to the
      pivotal row and to row 'col' are then interchanged. }
  var
    i, pivi : subs;
    rowp : rowptrs;
    max : real;
begin
  max := abs (a[col] ↑ .coeff[col]);
  pivi := col;
  for i := col+1 to n do
    if abs (a[i] ↑ .coeff[col]) > max then
    begin
      max := abs (a[i] ↑ .coeff[col]);   pivi := i
    end;
  if pivi <> col then
  begin
    rowp := a[col];   a[col] := a[pivi];   a[pivi] := rowp
  end
end { Swap Piv Row };


begin { Pointer Gauss }
  for j := 1 to n - 1 do
  begin
    SwapPivRow (eqn, j);
    pivotalrow := eqn[j];   pivot := pivotalrow ↑ .coeff[j];
    if abs (pivot) <= assumedzero then
    begin
      success := false;   goto 999
    end;
    for i := j + 1 to n do
    with eqn[i] ↑  do
    begin
      mult := coeff[j] / pivot;
      if abs (mult) > assumedzero then
```

```
            begin
               coeff[j] := mult;
               SubtractRow (eqn[i]↑ , pivotalrow↑ , mult, j + 1)
            end else
               coeff[j] := 0
         end
      end;
      success := abs(eqn[n]↑.coeff[n]) > assumedzero;
      if success then
      begin
         with eqn[n]↑ do
            x[n] := b / coeff[n];
         for i := n - 1 downto 1 do
         with eqn[i]↑ do
         begin
            top := b;
            for k := i + 1 to n do
               top := top - coeff[k] * x[k];
            x[i] := top / coeff[i]
         end
      end
   end { Pointer Gauss };
```

Figure 4.5

Unfortunately, not all programming languages provide pointers so that in many a vector of pointers must be simulated by a vector of row numbers. If pointers are not available, it is likely that records, too, are not available and the coefficient matrix will be represented as an $n \times n$ array. As this is a common situation, we illustrate the approach but stress that it is not to be recommended when using Pascal.

A permutation vector of type

array [*subs*] **of** *subs*

is constructed so that, initially,

$$p[i] = i \qquad \text{for } i = 1,2,\ldots,n$$

At the kth stage where, without pivotal interchanges, rows $k+1$, $k+2,\ldots,n$ would be worked through, the rows $p[k+1], p[k+2],\ldots,p[n]$ are used instead. When two rows $r_1 = p[i]$ and $r_2 = p[j]$ are to be interchanged, the two vector elements $p[i]$ and $p[j]$ are interchanged instead. The procedure of Fig. 4.6 illustrates the complete process. For our present purposes, the permutation vector is local to the procedure but it will become apparent in Section 4.1.4 that this aspect must be reconsidered.

```
procedure PermedGauss (var a : nbyn;   b : vectors;
                        var x : vectors;   var success : boolean);
  label 999;

  const
    assumedzero = 1E-20;

  var
    i, pi, pivi, j : subs;
    mult, pivot : real;
    p : array [subs] of subs;

  function dotprod (u,v : vectors;   k : subs) : real;
      { u[k. .n] . v[k. .n] }
    var
      i : subs;
      sum : real;
  begin
    sum := 0;
    for i := k to n do
      sum := sum + u[i] * v[i];
    dotprod := sum
  end { dot prod };

  procedure SubtractRow
      (var u : vectors;   v : vectors;   m : real;   k : subs);
        { u[k. .n] := u[k. .n] - m * v[k. .n] }
    var
      i : subs;
  begin
    for i := k to n do
      u[i] := u[i] - m * v[i]
  end { Subtract Row };

  procedure SwapPivRow
      (col : subs;   var pivrow : subs;   var p : permvectors);
        { Chooses, as the pivotal row, that row in
            p[col], p[col + 1], . . ., p[n]
          with the maximum element in column col
          and updates p to imply the interchange
          of row col and the pivotal row }
```

```
    var
      i, pivi : subs;
      max, thisabs : real;
  begin
    pivi := col;   max := abs (a[p[pivi], col]);
    for i := col + 1 to n do
    begin
      thisabs := abs (a[p[i], col]);
      if thisabs > max then
      begin
        max := thisabs;   pivi := i
      end
    end;
    pivrow := p[pivi];
      { now interchange p[col] and p[pivi] }
    if pivi <> col then
    begin
      i := p[col];   p[col] := p[pivi];   p[pivi] := i
    end
  end { Swap Piv Row };


begin { Permed Gauss }
  for i := 1 to n do p[i] := i;


  for j := 1 to n-1 do
  begin
    SwapPivRow (j, pivi, p);
    pivot := a[pivi, j];
    if abs (pivot) <= assumedzero then
    begin
      success := false;   goto 999
    end;
    for i := j + 1 to n do
    begin
      pi := p[i];
      mult := a[pi,j] / pivot;
      if abs (mult) > assumedzero then
      begin
        a[pi,j] := mult;
        SubtractRow (a[pi], a[pivi], mult, j+1);
        b[pi] := b[pi] - mult * b[pivi]
      end else
        a[pi,j] := 0
    end
  end;
```

```
    pi := p[n];
    success := abs(a[pi,n]) > assumedzero;
    if success then
    begin
      x[n] := b[pi] / a[pi,n];
      for i := n-1 downto 1 do
      begin
        pi := p[i];
        x[i] := (b[pi] - dotprod (a[pi], x, i + 1)) / a[pi,i]
      end
    end;
  999 :
  end { Permed Gauss };
```

Figure 4.6

4.1.3 Scaling

The pivoting strategy, described in Section 4.1.2, can be altered by rescaling the equations. For example, if we are given the system of equations

$$\begin{aligned} x_1 + \quad x_2 &= 2 \\ 0.0001x_1 + 0.1x_2 &= 1 \end{aligned} \tag{4.8}$$

then the multiplier is 0.0001. However, if we multiply the second equation through by 100 000 to give

$$\begin{aligned} x_1 + \qquad x_2 &= \qquad 2 \\ 10x_1 + 10000x_2 &= 100\,000 \end{aligned}$$

then we should interchange the rows and the multiplier becomes 0.1 instead. Hence, the pivoting strategy is affected by the scaling and, further, the calculations performed during the elimination are changed, giving a different accumulation of rounding errors. As yet, there is no agreement about a single 'best' way of scaling a system of equations to reduce the accumulation of rounding errors, but a method which is popular is that of 'equilibration'. In this technique the equations are scaled so that the maximum elements in the rows and columns of the matrix A have roughly the same magnitude, usually unity. An 'equilibration' of the equations (4.8) gives

$$\begin{aligned} x_1 + x_2 &= 2 \\ 0.001x_1 + x_2 &= 10 \end{aligned}$$

It is recommended that the equations be scaled before an elimination process is applied and equilibration is suggested as the technique to use for the first attempt to solve any system. For a more detailed discussion of scaling see Wilkinson (1965).

4.1.4 LU decomposition

It is interesting to rewrite the elimination phase of Gaussian elimination in terms of matrices. We shall start by assuming that pivoting is not necessary for the algorithm to proceed and, later, discuss the effect of its inclusion.

The first stage of the elimination process is to introduce zeros below the diagonal of the first column of $A_0=A$ to produce the matrix A_1. This is achieved by subtracting multiples

$$m_{i1} = \frac{a_{i1}}{a_{11}}$$

of the first row from the ith $\{i=2,3,...,n\}$. This can be written as the product

$$A_1 = L_1A_0$$

where L_1 is the unit lower triangular matrix

$$L_1 = \begin{bmatrix} 1 & 0 & \cdots & 0 \\ -m_{21} & 1 & \cdots & 0 \\ \cdot & \cdot & & \cdot \\ \cdot & \cdot & & \cdot \\ \cdot & \cdot & & \cdot \\ -m_{n1} & 0 & \cdots & 1 \end{bmatrix}$$

The algorithm proceeds by forming the product

$$A_2 = L_2A_1$$

of A_1 with the matrix

$$L_2 = \begin{bmatrix} 1 & 0 & 0 & \cdots & 0 \\ 0 & 1 & 0 & \cdots & 0 \\ 0 & -m_{32} & 1 & \cdots & 0 \\ \cdot & \cdot & \cdot & & \cdot \\ \cdot & \cdot & \cdot & & \cdot \\ \cdot & \cdot & \cdot & & \cdot \\ 0 & -m_{n2} & 0 & \cdots & 1 \end{bmatrix}$$

where the m_{i2} $\{i=3,4,...,n\}$ are the multipliers required to introduce zeros below the diagonal of the second column of A_1. In general, multiplying A_{i-1} by the matrix

$$L_i = \begin{bmatrix} 1 & 0 & \cdots & 0 & \cdots & 0 \\ 0 & 1 & \cdots & 0 & \cdots & 0 \\ \cdot & \cdot & & & & \cdot \\ \cdot & \cdot & & & & \cdot \\ \cdot & \cdot & & & & \cdot \\ 0 & 0 & \cdots & 1 & \cdots & 0 \\ 0 & 0 & \cdots & -m_{i+1,i} & \cdots & 0 \\ \cdot & \cdot & & & & \cdot \\ \cdot & \cdot & & & & \cdot \\ \cdot & \cdot & & & & \cdot \\ 0 & 0 & \cdots & -m_{ni} & \cdots & 1 \end{bmatrix}$$

to form the product

$$A_i = L_i A_{i-1} \qquad \text{for } i=1,2,\ldots,n-1$$

introduces zeros below the diagonal of the ith column of A_{i-1}. Thus

$$A_{n-1} = L_{n-1}L_{n-2}\cdots L_2 L_1 A \tag{4.9}$$

and A_{n-1} is an upper triangular matrix which we shall rename U.

Each of the matrices L_i $\{i=1,2,\ldots,n-1\}$ has the property that its inverse is L_i with the signs of the m_{ji} $\{j=i+1,\ldots,n\}$ reversed; thus

$$L_i^{-1} = \begin{bmatrix} 1 & 0 & \cdots & 0 & \cdots & 0 \\ 0 & 1 & \cdots & 0 & \cdots & 0 \\ \cdot & \cdot & & \cdot & & \cdot \\ \cdot & \cdot & & \cdot & & \cdot \\ \cdot & \cdot & & \cdot & & \cdot \\ 0 & 0 & \cdots & 1 & \cdots & 0 \\ 0 & 0 & \cdots & m_{i+1,i} & \cdots & 0 \\ \cdot & \cdot & & & & \cdot \\ \cdot & \cdot & & & & \cdot \\ \cdot & \cdot & & & & \cdot \\ 0 & 0 & \cdots & m_{ni} & \cdots & 1 \end{bmatrix}$$

The verification of this result is left as an exercise for the reader. Equation (4.9) can now be rewritten as

$$A = L_1^{-1}L_2^{-1}\cdots L_{n-2}^{-1}L_{n-1}^{-1}U$$

and it is left as an exercise to verify that the matrix

$$L = L_1^{-1}L_2^{-1}\cdots L_{n-2}^{-1}L_{n-1}^{-1}$$

takes the form

$$L = \begin{bmatrix} 1 & 0 & \cdots & 0 & \cdots & 0 \\ m_{21} & 1 & \cdots & 0 & \cdots & 0 \\ \cdot & \cdot & & \cdot & & \cdot \\ \cdot & \cdot & & \cdot & & \cdot \\ \cdot & \cdot & & \cdot & & \cdot \\ m_{i1} & m_{i2} & \cdots & 1 & \cdots & 0 \\ m_{i+1,1} & m_{i+1,2} & \cdots & m_{i+1,i} & \cdots & 0 \\ \cdot & \cdot & & \cdot & & \cdot \\ \cdot & \cdot & & \cdot & & \cdot \\ \cdot & \cdot & & \cdot & & \cdot \\ m_{n1} & m_{n2} & \cdots & m_{ni} & \cdots & 1 \end{bmatrix}$$

Thus

$$A = LU \tag{4.10}$$

The proofs of the following theorems can be found as exercises in Burden *et al.* (1981, Ex. 6.5.9 and Ex. 6.5.13).

Theorem 4.2

If Gaussian elimination, without pivoting, can be performed on the system of equations

$$A\mathbf{x} = \mathbf{b}$$

then A can be *factorised* or *decomposed* uniquely into the product

$$A = LU$$

of a lower triangular matrix L with unit diagonal and an upper triangular matrix U.

Definition 4.1

The $n \times n$ matrix A is said to be *strictly diagonally dominant* if

$$|a_{ii}| > \sum_{\substack{j=1 \\ j \neq i}}^{n} |a_{ij}| \qquad \text{for } i = 1,2,\ldots,n$$

Theorem 4.3

If A is strictly diagonally dominant then it is non-singular and Gaussian elimination can be performed on any system

$$Ax = \boldsymbol{b}$$

to obtain its unique solution without pivoting being necessary to control the growth of rounding errors.

We can see that this factorisation is useful by considering the solution of the system of equations

$$Ax = LUx = \boldsymbol{b}$$

This can be solved in two steps; first the system of equations

$$Ly = \boldsymbol{b} \tag{4.11}$$

is solved and then the solution to the system

$$Ux = y \tag{4.12}$$

is obtained. Hence x satisfies

$$L(Ux) = Ax = \boldsymbol{b}.$$

The solution of the upper triangular system (4.12) is achieved using the back-substitution technique already discussed and the solution of the system (4.11) involves a corresponding method of forward-substitution.

As an example, we consider the matrix

$$A = \begin{bmatrix} 2 & 1 & 3 \\ 4 & 3 & 10 \\ 2 & 4 & 17 \end{bmatrix}$$

of equation (4.5). We have already obtained the solution to (4.5) without pivoting and so, by Theorem 4.2, A has a unique LU decomposition. From (4.6) the multipliers used to introduce zeros into the first column of A are 2 and 1 and so

$$L_1 = \begin{bmatrix} 1 & 0 & 0 \\ -2 & 1 & 0 \\ -1 & 0 & 1 \end{bmatrix}$$

and

$$A_1 = L_1 A = \begin{bmatrix} 2 & 1 & 3 \\ 0 & 1 & 4 \\ 0 & 3 & 14 \end{bmatrix}$$

Further, the multiplier necessary to introduce the final zero into the second column of A_1 is 3 and so

$$L_2 = \begin{bmatrix} 1 & 0 & 0 \\ 0 & 1 & 0 \\ 0 & -3 & 1 \end{bmatrix}$$

giving

$$U = A_2 = L_2 A_1 = L_2 L_1 A = \begin{bmatrix} 2 & 1 & 3 \\ 0 & 1 & 4 \\ 0 & 0 & 2 \end{bmatrix} \tag{4.13a}$$

Finally

$$L = L_1^{-1} L_2^{-1} = \begin{bmatrix} 1 & 0 & 0 \\ 2 & 1 & 0 \\ 1 & 0 & 1 \end{bmatrix} \begin{bmatrix} 1 & 0 & 0 \\ 0 & 1 & 0 \\ 0 & 3 & 1 \end{bmatrix} = \begin{bmatrix} 1 & 0 & 0 \\ 2 & 1 & 0 \\ 1 & 3 & 1 \end{bmatrix} \tag{4.13b}$$

Thus solution of the system

$$L U \boldsymbol{x} = \boldsymbol{b} = \begin{bmatrix} 11 \\ 28 \\ 31 \end{bmatrix}$$

involves first applying forward-substitution to the system

$$\begin{bmatrix} 1 & 0 & 0 \\ 2 & 1 & 0 \\ 1 & 3 & 1 \end{bmatrix} \begin{bmatrix} y_1 \\ y_2 \\ y_3 \end{bmatrix} = \begin{bmatrix} 11 \\ 28 \\ 31 \end{bmatrix}$$

From the first equation

$$y_1 = 11$$

from the second

$$y_2 = 28 - 2(11) = 6$$

and from the third

$$y_3 = 31 - 1(11) - 3(6) = 2$$

Now back-substitution is applied to the system

$$\begin{bmatrix} 2 & 1 & 3 \\ 0 & 1 & 4 \\ 0 & 0 & 2 \end{bmatrix} \begin{bmatrix} x_1 \\ x_2 \\ x_3 \end{bmatrix} = \begin{bmatrix} 11 \\ 6 \\ 2 \end{bmatrix}$$

and gives the solution

$$x = \begin{bmatrix} 3 \\ 2 \\ 1 \end{bmatrix}$$

This technique is particularly useful if the same matrix A appears in conjunction with many different right-hand vectors $\boldsymbol{b}$. Once L and U have been determined, A can be discarded and the system

$$LU\boldsymbol{x} = \boldsymbol{b}$$

solved for each different $\boldsymbol{b}$. This, of course, is much faster than performing a full Gaussian elimination each time.

Because all the diagonal elements of L are unity they need not be stored and so both L and U can be stored in the space originally occupied by A. The two matrices of (4.13) would therefore be stored as the one matrix

$$\begin{bmatrix} 2 & 1 & 3 \\ 2 & 1 & 4 \\ 1 & 3 & 2 \end{bmatrix}$$

We recall that the procedure of Fig. 4.4 stores the matrices in this manner.

```
procedure LUsolve (row : rowvectors;   var x : vectors);

  var
    i, j : subs;

begin
  for i := 2 to n do
  with row[i] do
    for j := 1 to i-1 do
      b := b - coeff[j] * row[j].b;

  with row[n] do
    x[n] := b / coeff[n];

  for i := n - 1 downto 1 do
  with row[i] do
  begin
    for j := i + 1 to n do
      b := b - coeff[j] * x[j];
    x[i] := b / coeff[i]
  end
end { L U solve };
```

Figure 4.7

The procedure of Fig. 4.7 solves the system

$$LU\boldsymbol{x} = \boldsymbol{b}$$

on the assumption that L and U are stored together as a vector of rows. On a machine where **with** has no run-time effect, slicing should improve efficiency. This has not been done in Fig. 4.7.

There is an alternative approach to finding the LU decomposition, known as the *Doolittle* algorithm, which considers the analytic form of the product of the matrix L with the matrix U. For example, using the notation

$$L = \begin{bmatrix} 1 & 0 & \cdots & 0 \\ m_{21} & 1 & \cdots & 0 \\ \cdot & \cdot & & \cdot \\ \cdot & \cdot & & \cdot \\ \cdot & \cdot & & \cdot \\ m_{n1} & m_{n2} & \cdots & 1 \end{bmatrix}$$

and

$$U = \begin{bmatrix} u_{11} & u_{12} & \cdots & u_{1n} \\ 0 & u_{22} & \cdots & u_{2n} \\ \cdot & \cdot & & \cdot \\ \cdot & \cdot & & \cdot \\ \cdot & \cdot & & \cdot \\ 0 & 0 & \cdots & u_{nn} \end{bmatrix}$$

then

$$u_{1j} = a_{1j} \qquad \text{for } j = 1,2,\ldots,n$$

When u_{11} is known, m_{21} can be determined by observing that

$$m_{21}u_{11} = a_{21}$$

Once the top row of U has been determined and m_{21} is known, the second row of U can be computed from the following relations:

$$\begin{aligned} m_{21}u_{12} + u_{22} &= a_{22} \\ m_{21}u_{13} + u_{23} &= a_{23} \\ &\cdot \\ &\cdot \\ &\cdot \\ m_{21}u_{1n} + u_{2n} &= a_{2n} \end{aligned}$$

Continuing this process, the whole of L and U can be obtained.

In *Crout's* algorithm the matrix U, rather than L, is assumed to have the unit diagonal. Both of these algorithms have certain advantages over Gaussian elimination for producing LU factorisations whenever double precision arithmetic is available. As this is not generally the case using Pascal we shall not discuss these advantages further; instead we refer the reader to Atkinson (1978) for a more detailed discussion.

The LU decomposition takes a special form when A is symmetric and satisfies certain restrictions.

Definition 4.2

A *symmetric* $n \times n$ matrix A is said to be *positive definite* if

$$\boldsymbol{x}^{\mathrm{T}} A \boldsymbol{x} > 0$$

for every n-vector $\boldsymbol{x} \neq \boldsymbol{0}$. This condition is satisfied if *all* the eigenvalues of A are strictly positive.

Theorem 4.4

If A is a positive definite $n \times n$ matrix then it is non-singular and, further, Gaussian elimination can be performed on any system of equations

$$A\boldsymbol{x} = \boldsymbol{b}$$

to obtain its unique solution without pivoting being necessary to control the growth of rounding errors.

As a result of this theorem and of Theorem 4.2 we know that a positive definite matrix A has a unique LU decomposition. However, to reflect the symmetry of A, it is more usual to obtain a *symmetric* factorisation and the following theorem assures us that this is possible.

Theorem 4.5

If A is a positive definite $n \times n$ matrix then it has a factorisation of the form

$$A = LL^{\mathrm{T}}$$

where L is a lower triangular matrix.

This factorisation can be obtained in the same way as the LU decomposition except that, instead of enforcing $l_{ii}=1$ $\{i=1,2,\ldots,n\}$, the condition $l_{ii}=u_{ii}$ $\{i=1,2,\ldots,n\}$ is enforced. Further, if symmetry is fully taken into account, the factorisation requires about half as many multiplications and divisions $([n^3+9n^2+2n]/6)$.

Proofs of Theorems 4.4 and 4.5 can be found in Burden *et al* (1981, p.300 *et seq.*). The factorisation of A into the product LL^{T} is known as the *Choleski* factorisation of A.

Unfortunately, not all matrices have an LU factorisation. For example, the matrix

$$A = \begin{bmatrix} 0 & 1 \\ 1 & 1 \end{bmatrix} \tag{4.14}$$

does not have an LU factorisation. We have seen before that Gaussian elimination can be used to solve systems involving matrices of this type by exchanging (or permuting) rows of the matrix. This was given the name 'partial pivoting' and, in effect, corresponds to the premultiplication of the matrix A by a *permutation* matrix, P. Permutation matrices have the same form as the identity matrix except that the order of the rows has been altered. For example, premultiplication by the permutation matrix

$$P = \begin{bmatrix} 0 & 1 & 0 \\ 1 & 0 & 0 \\ 0 & 0 & 1 \end{bmatrix}$$

would interchange the first two rows of a 3×3 matrix. Premultiplication of A of (4.14) by the matrix

$$P = \begin{bmatrix} 0 & 1 \\ 1 & 0 \end{bmatrix}$$

gives

$$PA = \begin{bmatrix} 1 & 1 \\ 0 & 1 \end{bmatrix}$$

which *does* have an LU factorisation. Application of the elimination phase of Gaussian elimination with partial pivoting is equivalent to LU decomposition of a permuted version of the coefficient matrix. Premultiplication of the system

$$A\boldsymbol{x} = \boldsymbol{b}$$

by the permutation matrix P gives

$$PA\boldsymbol{x} = P\boldsymbol{b} = \boldsymbol{c}$$

and so the solution vector $\boldsymbol{x}$ can be found by forming the LU decomposition of PA and solving

$$LU\boldsymbol{x} = \boldsymbol{c}$$

by the standard technique.

Mention must be made of storing the factorised matrix. If the same coefficient matrix is to be associated with several right-hand sides, the elimination process should be performed once and the compacted *LU* matrix constructed and filed for future use. However, if partial pivoting has been employed, the rows of the *LU* matrix will be in some permuted order. When the *LU* matrix is used, this order must be known; a permutation vector indicating the order in which the rows are to be processed must be filed also. This vector is a compressed form of the permutation matrix above; its elements are integers recording the order in which rows of the identity matrix occur in the permutation matrix.

The *LU* matrix could be filed in its permuted form and then, when a new right-hand side vector, $\mathbf{y}$, is to be used, the program vector b would be constructed so that $b[i] = y_i$ and all references to rows of coefficients and to elements of b would be via the permutation vector.

A more sensible approach is to file the *LU* matrix in its 'proper' form with its rows in the order dictated by the permutation vector. Then, for a new right-hand vector $\mathbf{y}$, the program vector b is constructed so that

$$b[p[i]] = y_i$$

and the solution vector is obtained by direct application of the procedure of Fig. 4.7.

```
program PivotedGauss (lhs, rhs, lu, soln);

    { Applies Gaussian elimination with partial pivoting to a
      system of equations and outputs the solution and the LU
      factorisation of the coefficient matrix. Within the LU
      form, each row of coefficients is preceded by a row
      number to imply a permutation vector and is followed by
      a (redundant) right-hand side value. }

  const
    n = . . .;   assumedzero = . . .;

  type
    subs = 1 . . n;
    vectors = array [subs] of real;
    rowptrs = ↑ rows;
    rows = record
                rowno : subs;
                  coeff : vectors;
                      b : real
            end { rows };
    ptrvecs = array [subs] of rowptrs;
```

```
var
      lhs : file of vectors;
      rhs : file of real;
       lu : file of rows;
     soln : text;
        a : ptrvecs;
        x : vectors;
        i : subs;
     done : boolean;

  procedure PointerGauss
     (var eqn : ptrvecs;   var x : vectors;   var success : boolean);
  { as in Fig. 4.5 }
     . . .
  end {Pointer Gauss};

begin { program body }
   reset (lhs);   reset (rhs);
   for i := 1 to n do
   begin
      new (a[i]);
      with a[i] ↑ do
      begin
         rowno := i;   read (lhs, coeff);   read (rhs, b)
      end
   end;

   PointerGauss (a, x, done);

   rewrite (soln);
   if done then
   begin
      writeln (soln, 'System solved successfully');
      writeln (soln);
      for i := 1 to n do writeln (soln, x[i])
   end else
      writeln (soln, 'Zero pivot encountered');

   rewrite (lu);
   for i := 1 to n do write (lu, a[i] ↑ )
end.
```

Figure 4.8

If an explicit permutation vector is employed during the elimination phase, as in Fig. 4.6, its final form must be filed and so the procedure *PermedGauss* must be modified to make it available. There will be many run-time references to the permutation vector so it should remain a local variable and its final form assigned to a variable parameter. The procedure heading should be

```
procedure PermedGauss
  (var a : nbyn;   b : vectors;   var x : vectors;
   var pv : permvectors;   var success : boolean)
```

where *permvectors* is the type

```
array [subs] of subs
```

and the procedure body should conclude with the statement

```
pv := p
```

If the superior data representation of Fig. 4.5 is used, the permutation vector can easily be incorporated within the data structure. If each row explicitly records its own number, the permutation vector will automatically be saved when the *LU* matrix is filed. The program of Fig. 4.8 illustrates this. It assumes that the left-hand coefficients and the right-hand vector have been generated by program in two files *lhs* (a file of *vectors*) and *rhs* (a file of *real*). The *LU* form is written to a file *lu* (a file of *rows*) and the solution vector is written to a text file *soln*. If the file *rhs* is then updated, the program of Fig. 4.9 will solve the system for a new right-hand side.

```
program PivotedLU (lu, rhs, soln);

    { Accepts an LU factorisation in the form produced by
      the program PivotedGauss together with a right-hand
      vector b and solves the system LUx = b. }

  const
    n = . . .;
  type
    subs = 1 . . n;
    vectors = array [subs] of real;
```

```
  rows = record
           rowno : subs;
            coeff : vectors;
               b : real
         end { rows };
 rowvectors = array [subs] of rows;

 var
      lu : file of rows;
     rhs : file of real;
    soln : text;
       a : rowvectors;
       x,tempb : vectors;
          i,r : subs;

 procedure LUsolve (row : rowvectors;   var x : vectors);
   { as in Fig. 4.7 }
   . . .
 end { L U solve };

 begin { program body }
   reset (lu);
   for i := 1 to n do
     read (lu, a[i]);

   reset (rhs);
   for i := 1 to n do
     read (rhs, tempb[i]);
   for i := 1 to n do
   begin
     r := a[i].rowno;   a[r].b := tempb [r]
   end;
   LUsolve (a, x);

   rewrite (soln);
   for i := 1 to n do writeln (soln, x[i])
end.
```

Figure 4.9

4.1.5 The inverse matrix

The *inverse* of a real $n \times n$ matrix A is the $n \times n$ matrix A^{-1} with the property that

$$A^{-1}A = AA^{-1} = I \tag{4.15}$$

where I is the $n \times n$ identity matrix. It is rarely necessary to evaluate the inverse matrix because, in general, we can work with A itself and use an elimination process. Sometimes, however, the inverse matrix is required and then the LU factorisation, described in Section 4.1.4, can be used to evaluate it efficiently. From equation (4.15) we have

$$AA^{-1} = I$$

and so the product of A with the first column of A^{-1} gives the first column, $\boldsymbol{e}_1$, of the identity matrix. If the first column of A^{-1} is denoted by the vector $\boldsymbol{x}_1$, it can be obtained as the solution of the system of equations

$$A\boldsymbol{x}_1 = \boldsymbol{e}_1$$

In general, if $\boldsymbol{x}_j$ is the jth column of A^{-1} and $\boldsymbol{e}_j$ is the jth column of the identity matrix, then they satisfy the equation

$$A\boldsymbol{x}_j = \boldsymbol{e}_j \qquad \text{for } j = 1,2,\ldots,n. \tag{4.16}$$

Now (4.16) represents a situation in which the same matrix, A, appears with many right-hand sides, $\boldsymbol{e}_j$, so the solution vectors, $\boldsymbol{x}_j$, can be obtained easily using the LU factorisation of A. Hence, by solving the system

$$A\boldsymbol{x}_j = LU\boldsymbol{x}_j = \boldsymbol{e}_j \qquad \text{for } j = 1,2,\ldots,n$$

the inverse matrix is built up column by column to give

$$A^{-1} = [\boldsymbol{x}_1,\boldsymbol{x}_2,\ldots,\boldsymbol{x}_n]$$

Consequently, a program to calculate A^{-1} comprises two steps; firstly the LU factorisation of the matrix A is calculated and secondly the n systems of equations, represented by (4.16), are solved using this factorisation. Burden *et al.* (1981) calculate that the number of multiplications and divisions necessary to compute the inverse is

$$\frac{4n^3 - n}{3}$$

Some words of warning are necessary here. The calculated inverse should *never* be used to solve the system of equations

$$A\boldsymbol{x} = \boldsymbol{b}$$

by forming

$$\boldsymbol{x} = A^{-1}\boldsymbol{b}$$

because an unnecessary amount of computational effort, leading to increased rounding errors, would be required. Solution of the system

$$\boldsymbol{Ax} = \boldsymbol{b}$$

using LU decomposition involves one LU factorisation, one forward-substitution and one back-substitution. Construction of A^{-1} would involve the same LU factorisation plus n forward-substitutions and n back-substitutions and finally, to obtain $\boldsymbol{x}$, a matrix–vector multiplication would be necessary: altogether significantly more work. Furthermore, the matrix inverse is, in a sense, less accurate than the triangular factorisation because it is calculated from the factorisation and so yet more errors are introduced.

4.1.6 Solution of a tri-diagonal system

It is a feature of the solution techniques used for partial differential equations that strictly diagonally dominant (or positive definite) tri-diagonal systems of linear equations occur frequently. As with everything in numerical analysis we wish to take advantage of all the available information and, in this case, the structure of the matrix A is too specialised to be ignored.

Theorem 4.6

If A is a strictly diagonally dominant (or positive definite) tri-diagonal matrix, then it has a unique LU factorisation in which both L and U have only two diagonals: L the main and first sub-diagonal and U the main and first super-diagonal.

Proof

From Theorem 4.3 (or Theorem 4.4) A has a unique LU decomposition. The remainder of the proof is left as an exercise for the reader.

Consider the following tri-diagonal system of n equations

$$\begin{aligned}
d_1 x_1 + c_1 x_2 \qquad\qquad\qquad\qquad &= b_1 \\
a_2 x_1 + d_2 x_2 + c_2 x_3 \qquad\qquad &= b_2 \\
a_3 x_2 + d_3 x_3 + c_3 x_4 \qquad &= b_3 \\
&\ \ \vdots \\
a_{n-1} x_{n-2} + d_{n-1} x_{n-1} + c_{n-1} x_n &= b_{n-1} \\
a_n x_{n-1} + d_n x_n &= b_n
\end{aligned} \tag{4.17}$$

A saving in storage can be made by storing only the sub-, super- and main diagonals. Obviously, when n is large this saving is significant. The next thing to notice is that, in each column, there is only one non-zero element to be eliminated, resulting in a saving of computational effort.

If $d_1 \neq 0$, in (4.17), then x_1 can be eliminated from the second equation by forming the multiplier

$$m_1 = \frac{a_2}{d_1}$$

and producing the new equation

$$d_2'x_2 + c_2x_3 = b_2'$$

where

$$\begin{aligned} d_2' &= d_2 - m_1c_1 \\ b_2' &= b_2 - m_1b_1 \end{aligned}$$

Similarly, if $d_2' \neq 0$, x_2 can be eliminated from the third equation to give as the new third equation

$$d_3'x_3 + c_3x_4 = b_3'$$

where

$$m_2 = \frac{a_3}{d_2'}$$

and

$$\begin{aligned} d_3' &= d_3 - m_2c_2 \\ b_3' &= b_3 - m_2b_2' \end{aligned}$$

Continuing in this way, at the ith stage, x_i can be eliminated from equation $i+1$ (assuming $d_i' \neq 0$) to give the new equation

$$d_{i+1}'x_{i+1} + c_{i+1}x_{i+2} = b_{i+1}' \tag{4.18}$$

with

$$m_i = \frac{a_{i+1}}{d_i'} \tag{4.19}$$

and

$$d_{i+1}' = d_{i+1} - m_ic_i \tag{4.20}$$

$$b_{i+1}' = b_{i+1} - m_ib_i' \tag{4.21}$$

If the sequence (4.19), (4.20) and (4.21), resulting in (4.18), is repeated for $i = 1,2,\ldots,n-1$, then the original system is transformed to 'upper triangular' form and so the solution can be obtained using a modified version of the back-substitution discussed earlier. Assuming $d'_n \neq 0$ we have

$$x_n = \frac{b'_n}{d'_n}$$

and then, for $i = n-1, n-2, \ldots, 1$,

$$x_i = \frac{b'_i - c_i x_{i+1}}{d'_i}$$

```
procedure TriDiag (var eqn : rowvectors;   var x : vectors);

  var
    i : subs;
    pivot, mult, ci, bi : real;

begin { Tri Diag }
  for i := 1 to n - 1 do
  begin
    with eqn[i] do
    begin
      pivot := d;   ci := c;   bi := b
    end;
    with eqn[i + 1] do
    begin
      mult := a / pivot;
      if abs (mult) > assumedzero then
      begin
        a := mult;   d := d - mult * ci;   b :=  b - mult * bi
      end else
        a := 0
    end
  end;

  with eqn[n] do
    x[n] := b / d;
  for i := n - 1 downto 1 do
  with eqn[i] do
    x[i] := (b - c * x[i + 1]) / d
end { Tri Diag };
```

Figure 4.10

The procedure of Fig. 4.10 solves a tri-diagonal system assuming the data types

```
rows = record
        a, d, c, b : real
      end { rows };
rowvectors = array [subs] of rows;
```

in addition to *subs* and *vectors* of Section 1.3.3 (p.13).

It is likely that system (4.17) will need to be solved for many right-hand sides, so this procedure stores the multipliers m_k in the vector which originally contained the a_k.

4.1.7 Accuracy of the solution

If the above methods could be employed using exact arithmetic, then they would each produce the correct answer. However, as shown in Chapter 2, the fixed word length of computers leads to rounding errors and so the calculations performed in these methods are not exact. Consequently the effect of errors on the solution of the linear system must be considered.

If the *computed* solution of the system of linear equations is $\boldsymbol{x}^*$ and the *exact* solution is $\boldsymbol{x}$ then the 'size' of the error vector

$$\boldsymbol{e} = \boldsymbol{x} - \boldsymbol{x}^*$$

is of interest. Before we can discuss the behaviour of this error we must have some means of describing the magnitude of the error vector.

Norms

The 'size' of a vector is usually measured using what is known as a *norm*.

Definitions 4.3

The *norm* of a vector $\boldsymbol{x}$ is denoted by $\|\boldsymbol{x}\|$ and must satisfy the following conditions

(i) $\|\boldsymbol{x}\| \geq 0$ for all $\boldsymbol{x}$ and
$\|\boldsymbol{x}\| = 0$ if and only if $x_i = 0$ for all i
(ii) $\|c\boldsymbol{x}\| = |c| \, \|\boldsymbol{x}\|$ for any constant c,
(iii) $\|\boldsymbol{x}+\boldsymbol{y}\| \leq \|\boldsymbol{x}\| + \|\boldsymbol{y}\|$ (the triangle inequality).

The most commonly used norms are particular examples of the L_p norm.

Definition 4.4

The L_p-*norm*, $\|\boldsymbol{x}\|$, of the vector $\boldsymbol{x}$ is given by

$$\|\boldsymbol{x}\|_p = \left(\sum_{i=1}^{n} x_i{}^p \right)^{1/p}$$

Theorem 4.7

The L_p-norm satisfies the norm conditions.

Proof

The L_p-norm obviously satisfies (i) and (ii) above and, using Minkowski's inequality (Flett, 1966, p. 182), it can be seen that it also satisfies (iii).

If $p = 1$ then the L_1 norm of $\boldsymbol{x}$ is

$$\| \boldsymbol{x} \|_1 = \sum_{i=1}^{n} | x_i |$$

For $p = 2$

$$\| \boldsymbol{x} \|_2 = \left(\sum_{i=1}^{n} | x_i |^2 \right)^{1/2}$$

and in the special case, taking the limit as p tends to infinity,

$$\| \boldsymbol{x} \|_\infty = \max_{1 \le i \le n} [| x_i |] \quad \blacksquare$$

The L_∞ norm is sometimes known as the *maximum* or *uniform* norm. The L_2 vector norm corresponds to the Euclidean length of the vector and so is sometimes known as the *Euclidean* norm. It will be necessary to have a measure of the 'size' of a matrix and the vector norm can be used to define a corresponding matrix norm.

Definition 4.5

The norm of the matrix A *subordinate to* or *induced by* the vector norm $\| \boldsymbol{x} \|$ is defined by

$$\| A \| = \max_{\| \boldsymbol{x} \| = 1} (\| A\boldsymbol{x} \|)$$

for any of the vector norms described above. This norm satisfies (i), (ii) and (iii) above automatically and, for square matrices, the matrix norm must also satisfy

(iv) $\| AB \| \le \| A \| . \| B \|$

Theorem 4.8

The L_1-norm of the matrix A is given by

$$\| A \|_1 = \max_{1 \le j \le n} \left(\sum_{i=1}^{n} | a_{ij} | \right) \qquad \text{(column sum)}$$

and

Theorem 4.9

The L_∞-norm of the matrix A is given by

$$\| A \|_\infty = \max_{1 \le i \le n} \left(\sum_{j=1}^{n} | a_{ij} | \right) \qquad \text{(row sum)}$$

Proofs of Theorems 4.8 and 4.9 can be found in Burden *et al.* (1981). The L_2 matrix norm is slightly more difficult to derive.

Definition 4.6

The maximum modulus eigenvalue of the matrix A is denoted by $\rho(A)$ and is known as the *spectral radius* of A.

Theorem 4.10

The L_2-norm of the matrix A is given by

$$\| A \|_2 = (\rho(A^T A))^{1/2}$$

An outline of the proof of this result is given in Ralston and Rabinowitz (1978).

It follows immediately that if A is real and symmetric then

$$\| A \|_2 = \rho(A)$$

The norm

$$\| A \|_E = \left(\sum_{i=1}^{n} \sum_{j=1}^{n} | a_{ij} |^2 \right)^{1/2}$$

is known as the Euclidean norm of A and is *not* subordinate to the L_2 vector norm. Subordinate norms have the property that

$$\| I \| = 1$$

but for the Euclidean norm

$$\| I \|_E = n^{1/2}$$

Theorem 4.11

For any matrix norm

$$\rho(A) \leq \| A \|$$

Proof

If λ is an arbitrary eigenvalue of A and $\boldsymbol{x}$ its eigenvector, then

$$\lambda \boldsymbol{x} = A\boldsymbol{x}$$

Taking norms

$$\| \lambda \boldsymbol{x} \| = \| A\boldsymbol{x} \| \leq \| A \| \, \| \boldsymbol{x} \|$$

But

$$\| \lambda \boldsymbol{x} \| = | \lambda | \, \| \boldsymbol{x} \|$$

and so

$$| \lambda | \leq \| A \|$$

Now λ represents *any* eigenvalue of A and so

$$\rho(A) \leq \| A \|$$

It is less easy to prove the following result.

Theorem 4.12

There exist some matrix norm and an arbitrarily small $\epsilon > 0$ such that

$$\| A \| \leq \rho(A) + \epsilon$$

A proof of this can be found in Ortega (1972).

Error Analysis

We can now return to the effects of errors on the solution of the linear system. Initially we restrict ourselves to considering the effect, on the solution, of perturbations δA and $\delta \boldsymbol{b}$ in A and $\boldsymbol{b}$ respectively. These perturbations could be due, for example, to rounding errors incurred when the data was entered into the computer. First consider a perturbation, $\boldsymbol{\delta b}$, in $\boldsymbol{b}$ only; this induces a change, $\boldsymbol{\delta x}$, in $\boldsymbol{x}$ so that

$$A(\boldsymbol{x} + \boldsymbol{\delta x}) = \boldsymbol{b} + \boldsymbol{\delta b}$$

But

$$A\boldsymbol{x} = \boldsymbol{b}$$

so that $A\boldsymbol{\delta x}=\boldsymbol{\delta b}$ or $\boldsymbol{\delta x}=A^{-1}\boldsymbol{\delta b}$ and

$$\| \boldsymbol{\delta x} \| \leq \| A^{-1} \| \; \| \boldsymbol{\delta b} \|$$

giving a measure of the change in $\boldsymbol{x}$. The relative change $\| \boldsymbol{\delta x} \| / \| \boldsymbol{x} \|$ is often more useful. Now

$$\| \boldsymbol{b} \| \leq \| A \| \; \| \boldsymbol{x} \|$$

so that

$$\| \boldsymbol{b} \| \; \| \boldsymbol{\delta x} \| \leq \| A \| \; \| A^{-1} \| \; \| \boldsymbol{\delta b} \| \; \| \boldsymbol{x} \|$$

Division of both sides by $\| \boldsymbol{b} \| \; \| \boldsymbol{x} \|$ gives

Theorem 4.13

$$\frac{\| \boldsymbol{\delta x} \|}{\| \boldsymbol{x} \|} \leq \| A \| \; \| A^{-1} \| \; \frac{\| \boldsymbol{\delta b} \|}{\| \boldsymbol{b} \|}$$

which is an upper bound on the relative change in $\boldsymbol{x}$.

The term $\| A \| \; \| A^{-1} \|$ is obviously important. If $\| A \| \; \| A^{-1} \|$ is small then a small change in $\boldsymbol{b}$ will induce only a small change in $\boldsymbol{x}$; however, if $\| A \| \; \| A^{-1} \|$ is large then even a small change in $\boldsymbol{b}$ can induce a *large* change in $\boldsymbol{x}$.

Definition 4.7

The *condition number*, $K(A)$, of the matrix A is given by

$$K(A) = \| A \| \; \| A^{-1} \|$$

Theorem 4.14

$$K(A) \geq 1 \qquad \text{for all } A$$

Proof

$$1 = \| I \| = \| AA^{-1} \| \le \| A \| \ \| A^{-1} \| = K(A) \ \blacksquare$$

If $K(A)$ is very large then A is said to be *ill-conditioned*. Realistically, ill-conditioning should be discussed in terms of the accuracy of the arithmetic used during the computation. If t-digit arithmetic is being used (so that rounding errors are of the order of 10^{-t}) and if $K(A)$ is of the order of 10^s, then there is no point in calculating the solution to an accuracy of more than 10^{s-t}.

An expression for the relative change induced by a perturbation in A, alone, can also be written down. In this case, care is necessary since $A+\delta A$ can be singular even if A is not. This problem can be avoided by making the assumption that

$$\| A^{-1} \| \ \| \delta A \| < 1$$

Then, using the expression

$$A+\delta A = A(I + A^{-1}\delta A)$$

it can be shown that

Theorem 4.15

$$\frac{\| \delta \boldsymbol{x} \|}{\| \boldsymbol{x} \|} \le \frac{K(A)}{1 - K(A)\dfrac{\| \delta A \|}{\| A \|}} \frac{\| \delta \boldsymbol{b} \|}{\| \boldsymbol{b} \|}$$

Taking account of changes in both A and $\boldsymbol{b}$ and using the same assumption as above results in

Theorem 4.16

$$\frac{\| \delta \boldsymbol{x} \|}{\| \boldsymbol{x} \|} \le \frac{K(A)}{1 - K(A)\dfrac{\| \delta A \|}{\| A \|}} \left(\frac{\| \delta \boldsymbol{b} \|}{\| \boldsymbol{b} \|} + \frac{\| \delta A \|}{\| A \|} \right)$$

A proof of Theorem 4.15 (the proof of Theorem 4.16 is similar) appears in Ralston and Rabinowitz (1978). Again the condition number, $K(A)$, is all important.

The results presented here are useful in the sense that they point out a potential difficulty: the ill-conditioning of the matrix A. However, they take no account of the errors which occur during the computation of the solution of the system. The method of backward error analysis is designed to quantify the effects of such errors. In this approach the perturbation, E, for which the computed solution, $\boldsymbol{x}^*$, satisfies the equation

$$(A + E)\,\boldsymbol{x}^* = \boldsymbol{b}$$

exactly, is determined. Now, by bounding E, a measure of the accuracy of the solution can be obtained. Deriving the results obtained using backward error analysis is beyond the scope of this book but two results are worth quoting; Wilkinson (1965) has proved the following two theorems.

Theorem 4.17

If t-digit arithmetic is used then the perturbation, E, induced by Gaussian elimination with pivoting satisfies the inequality

$$\| E \|_\infty \leq f(n)\, 10^{1-t} \max_{i,j,k} | a_{ij}^{(k)} |$$

where $a_{ij}^{(k)}$ is the (i,j)th element of the reduced form of A at the kth stage in the elimination.

Theorem 4.18

In practice

$$f(n) \simeq n$$

but, at worst,

$$f(n) \leq 1.01(n^3 + 3n^2)$$

Thus if A is well-conditioned so that the $a_{ij}^{(k)}$ do not become large and provided that n is not large, then, effectively, this result says that the accumulated rounding errors are of the same order as those incurred when representing the elements of A in floating-point form.

4.2 Iterative methods

Iterative methods are particularly useful when the system of equations is large and most of the coefficients are zero. Systems of this type occur frequently in the solution of problems involving partial differention equations and it is the need for such solutions that has prompted much of the research into these methods.

The basic idea behind the methods in this section is essentially the same as that in Section 3.1.4 in which the equation

$$f(x) = 0$$

was rearranged to give an iteration formula

$$x_{k+1} = g(x_k)$$

for finding a zero of $f(x)$. Here, there are n equations

$$\begin{aligned} a_{11}x_1 + a_{12}x_2 + \cdots + a_{1n}x_n &= b_1 \\ a_{21}x_1 + a_{22}x_2 + \cdots + a_{2n}x_n &= b_2 \\ &\vdots \\ a_{n1}x_1 + a_{n2}x_2 + \cdots + a_{nn}x_n &= b_n \end{aligned} \tag{4.22}$$

rather than one.

4.2.1 Jacobi's method

The structure of the equations suggests an obvious rearrangement. Consider the first equation and let us suppose that, from it, we want an iteration formula which will give a new value to the variable x_1. The simplest way to achieve this is to take every term in the equation, except that involving x_1, over to the right-hand side of the equation and divide through by a_{11} to give

$$x_1 = \frac{b_1 - a_{12}x_2 - a_{13}x_3 - \cdots - a_{1n}x_n}{a_{11}}$$

which leads to the iteration

$$x_1^{(k+1)} = \frac{b_1 - a_{12}x_2^{(k)} - \cdots - a_{1n}x_n^{(k)}}{a_{11}}$$

Similarly the second equation can be rearranged to give an iteration for x_2 and, in general, an iteration for x_i can be obtained from the ith equation

$$x_i^{(k+1)} = \frac{b_i - a_{i1}x_1^{(k)} - \cdots - a_{ii-1}x_{i-1}^{(k)} - a_{i,i+1}x_{i+1}^{(k)} - \cdots - a_{in}x_n^{(k)}}{a_{ii}} \tag{4.23}$$

Hence, following on from the methods of Chapter 3, an algorithm for finding the solution to the system of equations (4.22) suggests itself.

(i) Choose an initial approximation, $\boldsymbol{x}^{(0)}$, to the solution. If no information about the solution is available, then choose

$$x_i^{(0)} = \frac{b_i}{a_{ii}}, \qquad i = 1,2,\ldots,n.$$

(ii) Determine the next iterate $\boldsymbol{x}^{(k+1)}$ from $\boldsymbol{x}^{(k)}$, $k = 0,1,\ldots$, using equation (4.23) for $i = 1,2,\ldots,n$.

(iii) If the modulus of *each* of the differences

$$x_i^{(k+1)} - x_i^{(k)}, \qquad i = 1,2,\ldots,n$$

is less than some prescribed tolerance, then stop the iteration and take the latest iterate $\boldsymbol{x}^{(k+1)}$ as the solution. Otherwise return to step (ii) replacing k by $k+1$.

This process is known as the *Jacobi* iteration and is illustrated by the procedure of Fig. 4.11. A discussion of the convergence properties of this method is deferred until Section 4.2.4.

```
procedure Jacobi
    (a : nbyn;   b : vectors;   var x : vectors;
     tol : real;   maxits : posints;
     var converged : boolean;   var noofits : posints);
  var
    itcount : 0 . . maxint;
    allwithintol : boolean;
    i, j : subs;
    rhs : real;
    oldx : vectors;
begin { Jacobi }
  itcount := 0;
  repeat
    oldx := x;   allwithintol := true;
    itcount := itcount + 1;
    for i := 1 to n do
    begin
      rhs := b[i];
      for j := 1 to i − 1 do rhs := rhs − a[i,j] * oldx[j];
      for j := i + 1 to n do rhs := rhs − a[i,j] * oldx[j];
      rhs := rhs/a[i,i];   x[i] := rhs;
      if abs(rhs − oldx[i]) > tol then
        allwithintol := false
    end
  until allwithintol or (itcount = maxits);
  converged := allwithintol;   noofits := itcount
end { Jacobi };
```

Figure 4.11

So that the Pascal procedure directly reflects the mathematics, the coefficient matrix A has been represented by a two-dimensional array, $\boldsymbol{x}$ and $\boldsymbol{b}$ are both vectors and no slicing is performed. A better data representation will be suggested in Section 4.2.2.

Notice how the convergence test is implemented. At the start of each iteration a boolean variable *allwithintol* is set *true* and then, during the iteration, it becomes *false* if any newly computed element of $\boldsymbol{x}$ differs from its previous value by more than the specified tolerance.

4.2.2 The Gauss–Seidel method

As with a single non-linear equation, the question 'Has all the available information been used?' can be asked. The answer is 'No'; when equation (4.23) is being used to calculate a new iterate, $x_i^{(k+1)}$, new values $x_1^{(k+1)}$, $x_2^{(k+1)}, \ldots, x_{i-1}^{(k+1)}$, have already been calculated. Therefore, in order to use this latest information, the iteration of equation (4.23) can be modified to give the form

$$x_i^{(k+1)} = \frac{b_i - a_{i1}x_1^{(k+1)} - \cdots - a_{i,i-1}x_{i-1}^{(k+1)} - a_{i,i+1}x_{i+1}^{(k)} - \cdots - a_{in}x_n^{(k)}}{a_{ii}} \quad (4.24)$$

When $i = 1$ the right-hand side includes terms with superscript (k) only and when $i = n$, terms with superscript $(k+1)$ only. The iteration (4.24), applied for $i = 1, 2, \ldots, n$, is known as the *Gauss–Seidel* iteration and, in general, the iterates converge more rapidly than do those of Jacobi's method although, for certain examples, the initial convergence of Jacobi's method can be faster. A further discussion of the rates of convergence is given in Section 4.2.4.

Within the Gauss–Seidel process, operations on the coefficient matrix are row-wise and for each row one element of $\boldsymbol{b}$ is involved but several elements of $\boldsymbol{x}$ are needed. This was the case for both Jacobi iteration and Gaussian elimination. Consequently, the most appropriate data representations within programs for Jacobi and Gauss–Seidel are those best suited to Gaussian elimination. The solution vector $\boldsymbol{x}$ is represented as a *vector* but the coefficient matrix A and the right-hand vector $\boldsymbol{b}$ are combined into a vector of records. Each record contains one row of coefficients and the corresponding element of $\boldsymbol{b}$. Some saving in storage can be achieved by noting that, once a new value $x_i^{(k+1)}$ has been calculated and the convergence tested by comparing it with $x_i^{(k)}$, the old value $x_i^{(k)}$ is no longer needed and can be overwritten. The procedure of Fig. 4.12 applies the Gauss–Seidel iteration and assumes the recommended data representations.

On a machine where **with** has no run-time effect, efficiency would be improved by slicing. The with-statement would then have the form

```
with eqn [i] do
begin
  newx := (b - prod (coeff, x, i)) / coeff[i];
  if abs(newx - x[i]) > tol then
    allwithintol := false;
  x[i] := newx
end
```

using the following function

```
function prod (var u, v : vectors;   diag : subs) : real;
  var
    i : subs;
    sum : real;
begin
  sum := 0;
  for i := 1 to diag - 1 do sum := sum + u[i] * v[i];
  for i := diag + 1 to n do sum := sum + u[i] * v[i];
  prod := sum
end { prod };
```

4.2.3 Successive over-relaxation

Further attempts have been made to increase the rate of convergence of the Jacobi and Gauss–Seidel iterations. It is possible, for example, to use Aitken's Δ^2 process to try to speed up the progress of the iteration. One of the best known methods, and the only one considered here, is that of *successive over-relaxation*, commonly known as SOR. This method is based on the idea that having produced a new value $\bar{x}_i^{(k+1)}$, using the Gauss–Seidel method, an even better value may result by forming a weighted average of the old and new values. Hence, the 'final' value, $x_i^{(k+1)}$, is given by

$$x_i^{(k+1)} = \omega\bar{x}_i^{(k+1)} + (1-\omega)x_i^{(k)} \tag{4.25}$$

where ω (>0) is an arbitrary parameter independent of k and i. Notice that, if $\omega = 1$, then $x_i^{(k+1)}$ is the value given by the Gauss–Seidel method. If $\omega > 1$ then (4.25) defines the method of successive over-relaxation and, for a large class of matrices, it can be shown that the optimum value of ω lies between 1 and 2 (see Theorems 4.21 and 4.22). This optimum ω is difficult to determine and, usually, an educated guess must suffice (but see Theorem 4.22 for a special case).

The effect of SOR can be demonstrated with the system

$$A = \begin{bmatrix} 10 & -8 & 0 \\ -8 & 10 & -1 \\ & & 10 \end{bmatrix} \qquad \boldsymbol{b} = \begin{bmatrix} -6 \\ 9 \\ 28 \end{bmatrix}$$

which has solution [1,2,3]. Using the initial approximation $x_i = b_i/a_{ii}$ $\{i = 1,2,\ldots,n\}$ and a tolerance of 5E−7, Gauss–Seidel converges after 33 iterations and Jacobi requires 65. For this coefficient matrix, the optimal value of ω can be calculated and, to four decimal places, it is 1.2566. Using this value, and with the same initial approximation and tolerance, SOR produces convergence after only 14 iterations.

```
procedure GaussSeidel
     (eqn : rowvectors;   var x : vectors;
      tol : real;   maxits : posints;
      var converged : boolean;   var noofits : posints);
  var
     itcount : 0 . . maxint;
     i, j : subs;
     allwithintol : boolean;
     newx : real;
begin
  itcount := 0;
  repeat
     allwithintol := true;   itcount := itcount + 1;
     for i := 1 to n do
     with eqn[i] do
     begin
        newx := b;
        for j := 1 to i-1 do newx := newx - coeff[j] * x[j];
        for j := i+1 to n do newx := newx - coeff[j] * x[j];
        newx := newx / coeff[i];
        if abs(newx - x[i]) > tol then
           allwithintol := false;
        x[i] := newx
     end
  until allwithintol or (itcount = maxits);
  converged := allwithintol;   noofits := itcount
end { Gauss Seidel };
```

Figure 4.12

4.2.4 Convergence

The three methods discussed in this section are examples of the general iterative scheme

$$x^{(k+1)} = Bx^{(k)} + c \tag{4.26}$$

The matrix A can be written in the form

$$A = L + D + U$$

where D is a non-zero diagonal matrix, L is strictly lower triangular and U is strictly upper triangular. If A is non-singular, then such a D can always be found by reordering the rows and columns of A. Using this notation the Jacobi iteration matrix is

$$B_J = -D^{-1}(L + U)$$

and

$$\boldsymbol{c}_J = D^{-1}\boldsymbol{b}$$

the Gauss–Seidel iteration matrix is

$$B_{GS} = -(D + L)^{-1}U$$

with

$$\boldsymbol{c}_{GS} = (D + L)^{-1}\boldsymbol{b}$$

and for SOR

$$B_{SOR} = (D + \omega L)^{-1}((1-\omega)D - \omega U)$$

and

$$\boldsymbol{c}_{SOR} = \omega(D + \omega L)^{-1}\boldsymbol{b}$$

The verification of these results is left as an exercise for the reader.

Definition 4.8

The iterates produced using (4.24) are said to converge to the vector $\boldsymbol{x}$, as k tends to infinity, if

$$\lim_{k\to\infty} \boldsymbol{x}^{(k)} = \boldsymbol{x}$$

or, equivalently,

$$\lim_{k\to\infty} (\boldsymbol{x} - \boldsymbol{x}^{(k)}) = \boldsymbol{0}$$

Theorem 4.19

The iteration

$$x^{(k+1)} = Bx^{(k)} + c$$

will converge in the norm $\| \cdot \|$ if

$$\| B \| < 1$$

Proof

Taking limits, as $k \to \infty$, of both sides of (4.26) it can be seen that x must satisfy

$$x = Bx + c$$

The error, $e^{(k+1)}$, in the $(k+1)$st iterate, $x^{(k+1)}$, satisfies

$$\begin{aligned} e^{(k+1)} &= x - x^{(k+1)} \\ &= (Bx + c) - (Bx^{(k)} + c) \\ &= B(x - x^{(k)}) \\ &= Be^{(k)} \end{aligned}$$

Application of this result again gives

$$e^{(k+1)} = B^2 e^{(k-1)}$$

and, finally,

$$e^{(k+1)} = B^{k+1} e^{(0)}$$

where $e^{(0)}$ is the error in the initial approximation, $x^{(0)}$, to x. Now the application of norms results in

$$\begin{aligned} \| e^{(k+1)} \| &= \| B^{k+1} e^{(0)} \| \\ &\leq \| B \|^{k+1} \| e^{(0)} \| \end{aligned}$$

and the iteration is convergent if

$$\| B \| < 1 \blacksquare$$

As a result of Theorem 4.12 it follows that if the spectral radius of B is strictly less than 1 then there exist a norm and an arbitrarily small $\epsilon > 0$ such that

$$\| B \| \leq \rho(b) + \epsilon < 1$$

and the iterative method will be convergent in that norm.

It is possible to prove the following theorem concerning the convergence of the Jacobi and Gauss–Seidel methods for matrices with non-negative elements.

Theorem 4.20

One of the following conditions holds

(i) $\rho(B_J) = \rho(B_{GS}) = 0$
(ii) $\rho(B_J) = \rho(B_{GS}) = 1$
(iii) $0 < \rho(B_{GS}) < \rho(B_J) < 1$
(iv) $1 < \rho(B_J) < \rho(B_{GS})$

A proof of this theorem, known as the Stein–Rosenberg theorem, can be found in Young (1971).

Following Theorem 4.20 the Jacobi and Gauss–Seidel iterations both converge or both diverge, and when they both converge, the Gauss–Seidel method converges faster (except in the trivial case, (i)). Further, for the SOR iteration it is possible to prove

Theorem 4.21

$\rho(B_{SOR}) < 1$ only if $0 < \omega < 2$

and

Theorem 4.22

If A is positive definite, then

$$\rho(B_{GS}) = (\rho(B_J))^2 < 1$$

and the optimal SOR parameter, ω, is given by

$$\omega_{opt} = \frac{2}{1 + (1 - (\rho(B_J)^2))^{1/2}}$$

and, with this choice, $\rho(B_{SOR}) = \omega_{opt} - 1$.

The calculation of the spectral radii of the iteration matrices, to check convergence, is not simple and so it is desirable to have a convergence test which is more easily verified. If the matrix A is strictly diagonally dominant, then the (i,j)th element of B_J is

$$(B_J)_{ij} = \begin{cases} \dfrac{a_{ij}}{a_{ii}} & i \neq j \\ 0 & i = j \end{cases}$$

and so

$$\| B_J \|_\infty = \max_i \left(\sum_{\substack{j=1 \\ j \neq i}}^{n} \frac{| a_{ij} |}{| a_{ii} |} \right) < 1$$

Hence the Jacobi method converges in the L_∞ norm for strictly diagonally dominant matrices, which are easily identified. The same result holds for the Gauss–Seidel method but is more difficult to prove. Testing for diagonal dominance is more restrictive than checking the spectral radius, with the result that both methods may still converge for matrices which are not strictly diagonally dominant.

4.3 Exercises

1 Prove the result quoted in Theorem 4.1.

2 (i) The system of equations

$$\begin{aligned} 2x_1 + 3x_2 + x_3 &= 2 \\ x_1 + x_2 + 2x_3 &= 4 \\ 3x_1 + 4x_2 + 3x_3 &= 6 \end{aligned}$$

is singular. By running a program, verify that the procedure of Fig. 4.2 will identify the singularity.

(ii) Verify that the system of equations

$$\begin{aligned} 2x_1 + 3x_2 + x_3 &= 2 \\ x_1 + x_2 + 2x_3 &= 4 \\ 3x_1 + 4x_2 + 3x_3 &= 7 \end{aligned}$$

does not have a solution by attempting to solve the system by hand using Gaussian elimination with partial pivoting. Modify the procedure of Fig. 4.6 so that it can differentiate between singularity and the lack of a solution and test your modification by applying it to solve the above system.

continued

3 Verify that the inverse of the matrix

$$L_i = \begin{bmatrix} 1 & 0 & \cdots & 0 & \cdots & 0 \\ 0 & 1 & \cdots & 0 & \cdots & 0 \\ \cdot & \cdot & & \cdot & & \cdot \\ \cdot & \cdot & & \cdot & & \cdot \\ \cdot & \cdot & & \cdot & & \cdot \\ 0 & 0 & \cdots & 1 & \cdots & 0 \\ 0 & 0 & \cdots & -m_{i+1,i} & \cdots & 0 \\ \cdot & \cdot & & \cdot & & \cdot \\ \cdot & \cdot & & \cdot & & \cdot \\ \cdot & \cdot & & \cdot & & \cdot \\ 0 & 0 & \cdots & -m_{ni} & \cdots & 1 \end{bmatrix}$$

is

$$L_i^{-1} = \begin{bmatrix} 1 & 0 & \cdots & 0 & \cdots & 0 \\ 0 & 1 & \cdots & 0 & \cdots & 0 \\ \cdot & \cdot & & \cdot & & \cdot \\ \cdot & \cdot & & \cdot & & \cdot \\ \cdot & \cdot & & \cdot & & \cdot \\ 0 & 0 & \cdots & 1 & \cdots & 0 \\ 0 & 0 & \cdots & m_{i+1,i} & \cdots & 0 \\ \cdot & \cdot & & \cdot & & \cdot \\ \cdot & \cdot & & \cdot & & \cdot \\ \cdot & \cdot & & \cdot & & \cdot \\ 0 & 0 & \cdots & m_{ni} & \cdots & 1 \end{bmatrix}$$

for $i=1,2,\ldots,n$.

4 Verify that the product

$$L = L_1^{-1} L_2^{-1} \cdots L_{n-2}^{-1} L_{n-1}^{-1}$$

takes the form

$$L = \begin{bmatrix} 1 & 0 & \cdots & 0 & \cdots & 0 \\ m_{21} & 1 & \cdots & 0 & \cdots & 0 \\ \cdot & \cdot & & \cdot & & \cdot \\ \cdot & \cdot & & \cdot & & \cdot \\ \cdot & \cdot & & \cdot & & \cdot \\ m_{i1} & m_{i2} & \cdots & 1 & \cdots & 0 \\ m_{i+1,1} & m_{i+1,2} & \cdots & m_{i+1,i} & \cdots & 0 \\ \cdot & \cdot & & \cdot & & \cdot \\ \cdot & \cdot & & \cdot & & \cdot \\ \cdot & \cdot & & \cdot & & \cdot \\ m_{n1} & m_{n2} & \cdots & m_{ni} & \cdots & 1 \end{bmatrix}$$

5 Write a program to use the Doolittle algorithm of Section 4.1.4 to find the L and U factors of a matrix A. Apply your program to find the L and U factors of the matrix

$$A = \begin{bmatrix} 2 & 1 & 3 \\ 4 & 3 & 10 \\ 2 & 4 & 17 \end{bmatrix}$$

[Note: L and U are given in Section 4.1.4.]

6 Show that a permutation matrix, P, is its own inverse. Hence describe how the LU decomposition technique, with partial pivoting, can be used to find the inverse of a matrix A.

7 Prove the result of Theorem 4.6.

8 Use the procedure *TriDiag*, of Fig. 4.10, to solve the system of equations

$$\begin{bmatrix} 4 & -1 & 0 & 0 & 0 \\ -1 & 4 & -1 & 0 & 0 \\ 0 & -1 & 4 & -1 & 0 \\ 0 & 0 & -1 & 4 & -1 \\ 0 & 0 & 0 & -1 & 4 \end{bmatrix} \boldsymbol{x} = \begin{bmatrix} 100 \\ 200 \\ 200 \\ 200 \\ 100 \end{bmatrix}$$

This system of equations is typical of those encountered in the numerical solution of partial differential equations.

9 The *Hilbert* matrix is classically ill-conditioned. The 4×4 Hilbert matrix is

$$H_4 = \begin{bmatrix} 1 & \frac{1}{2} & \frac{1}{3} & \frac{1}{4} \\ \frac{1}{2} & \frac{1}{3} & \frac{1}{4} & \frac{1}{5} \\ \frac{1}{3} & \frac{1}{4} & \frac{1}{5} & \frac{1}{6} \\ \frac{1}{4} & \frac{1}{5} & \frac{1}{6} & \frac{1}{7} \end{bmatrix}$$

For the system $H_4\boldsymbol{x}=\boldsymbol{b}$, with $\boldsymbol{b}^T=(4, \frac{163}{60}, \frac{21}{10}, \frac{241}{140})$, the solution is $\boldsymbol{x}=(1,2,3,4)^T$. Use the procedures

(i) *Gauss* (Fig. 4.1) and
(ii) *PointerGauss* (Fig. 4.5)

to obtain a solution to the system and compare your answers.

10 Scale the system of equations in Exercise 9 and solve the scaled system using *PointerGauss*. Compare your answers with those in Exercise 9.

11 Calculate the inverse of the 4×4 Hilbert matrix and hence solve the system in Exercise 9, comparing your answer with those in Exercises 9 and 10.

12 Prove that if there are perturbations δA and $\boldsymbol{\delta b}$ in A and $\boldsymbol{b}$, respectively, then the perturbation $\boldsymbol{\delta x}$ induced in the solution of the equation

$$A\boldsymbol{x} = \boldsymbol{b}$$

satisfies the inequality

$$\frac{\| \boldsymbol{\delta x} \|}{\| \boldsymbol{x} \|} \leq \frac{K(A)}{1 - K(A)\dfrac{\| \delta A \|}{\| A \|}} \left(\frac{\| \boldsymbol{\delta b} \|}{\| \boldsymbol{b} \|} + \frac{\| \delta A \|}{\| A \|} \right)$$

13 Use the procedure *GaussSeidel* (Fig. 4.12) to solve the non-strictly diagonally dominant system of equations

$$\begin{aligned} -5x_1 + 2x_2 + 2x_3 &= -8 \\ 3x_1 - 7x_2 + 2x_3 &= -7 \\ 3x_1 + 4x_2 - 7x_3 &= -11 \end{aligned}$$

correct to four decimal places.

14 Write a program which uses the method of successive over-relaxation to solve a system of equations $A\boldsymbol{x}=\boldsymbol{b}$. The program should include the facility for changing the acceleration parameter, ω, as the iteration progresses. Use the program to solve the system

$$\begin{aligned} 4x_1 + 3x_2 \phantom{{}-x_3} &= 24 \\ 3x_1 + 4x_2 - x_3 &= 30 \\ -x_2 + 4x_3 &= -24 \end{aligned}$$

Compare your results, using your choice of ω, with those obtained using $\omega=1.25$ throughout.

Chapter 5 EIGENVALUES AND EIGENVECTORS

Eigenvalues are very important in obtaining a full understanding of many physical systems. For example, the stability of aircraft in flight and the modes of vibration of bridges are governed by eigenvalues. They are also important in the study of the growth of certain types of population. The population can be represented by a vector $\boldsymbol{y}_t$, each component of which represents the number of females of a certain age in that population; the subscript t represents time. Leslie (1948) derived a model for the birth/death process in the population and it takes the form

$$\boldsymbol{y}_{t+1} = A\boldsymbol{y}_t$$

where A is a square matrix. It is known that if the population is in the state of age stability, then there is a positive real number λ such that

$$\boldsymbol{y}_{t+1} = \lambda \boldsymbol{y}_t$$

and so

$$A\boldsymbol{y}_t = \lambda \boldsymbol{y}_t$$

The constant λ is known as the finite rate of increase of the population and determines whether or not the population increases; if $\lambda>1$ then the total population increases, if $\lambda=1$ it remains the same and if $\lambda<1$ it decreases. Now λ is an *eigenvalue* of the matrix A and $\boldsymbol{y}_t$ is an *eigenvector* of A corresponding to the eigenvalue λ. A more formal definition of eigenvalues and eigenvectors follows.

Definition 5.1

If A is an $n \times n$ real matrix then its n *eigenvalues* $\lambda_1, \lambda_2, \ldots, \lambda_n$ are the real and complex roots of the *characteristic polynomial*

$$p(\lambda) = \det(A - \lambda I)$$

If λ is an eigenvalue of A and the vector $\boldsymbol{x} \neq \boldsymbol{0}$ has the property that

$$A\boldsymbol{x} = \lambda \boldsymbol{x} \tag{5.1}$$

or

$$(A - \lambda I)\boldsymbol{x} = \boldsymbol{0} \tag{5.2}$$

then $\boldsymbol{x}$ is called an *eigenvector* of A corresponding to the eigenvalue λ.

The vector $\boldsymbol{x}$ in (5.1) is known as a *right* eigenvector of A corresponding to λ. There also exists a *left* eigenvector $\boldsymbol{y}$ such that

$$\boldsymbol{y}^{\mathrm{T}}A = \lambda \boldsymbol{y}^{\mathrm{T}}$$

If A is real and symmetric, then

$$(A\boldsymbol{x})^{\mathrm{T}} = \boldsymbol{x}^{\mathrm{T}}A^{\mathrm{T}} = \boldsymbol{x}^{\mathrm{T}}A$$

and

$$(\lambda \boldsymbol{x})^{\mathrm{T}} = \lambda \boldsymbol{x}^{\mathrm{T}}$$

and so $\boldsymbol{x}$ is both a right and a left eigenvector. In the material which follows only right eigenvectors will be considered and so the prefix 'right' will be dropped.

An eigenvector $\boldsymbol{x}$ is unique only to a constant multiple; the vector $\alpha\boldsymbol{x}$, for α constant, is also an eigenvector of A corresponding to λ since

$$A(\alpha\boldsymbol{x}) = \alpha(A\boldsymbol{x}) = \alpha(\lambda\boldsymbol{x}) = \lambda(\alpha\boldsymbol{x})$$

To remove this ambiguity it is usual to *normalise* the eigenvector in one of two ways by insisting that either

$$\|\boldsymbol{x}\|_\infty = \max_i (\,|x_i|\,) = 1$$

or

$$\|\boldsymbol{x}\|_2 = \left(\sum_{i=1}^{n} |x_i|^2\right)^{1/2} = 1$$

Section 5.4 contains a short discussion of non-symmetric systems but, to simplify the analysis, we concentrate on finding the eigenvalues and eigenvectors of real symmetric matrices. In the following section various definitions and theorems, necessary for the study of eigenvalues, are quoted. Proofs are either left as exercises or can be found in any standard text on linear algebra, for example Gourlay and Watson (1973) or Wilkinson (1965).

5.1 Definitions and theorems

Definition 5.2

A set of vectors $\{y_1, y_2, \ldots, y_n\}$ is said to be *orthogonal* if

$$y_i^T y_j = 0$$

for all $i \neq j$. If, in addition, $y_i^T y_i = 1$, for all $i = 1, 2, \ldots, n$, then the set is said to be *orthonormal*. An $n \times n$ matrix P is said to be orthogonal if

$$P^T P = I$$

Note that this implies

$$P^{-1} = P^T$$

Theorem 5.1

The eigenvalues of a symmetric matrix are real.

A result which can be proved quite simply is

Theorem 5.2

Eigenvectors corresponding to distinct eigenvalues of a symmetric matrix are orthogonal.

Theorem 5.3

If A is a symmetric matrix then there exists an orthogonal matrix P such that

$$P^{-1}AP = D$$

where D is a diagonal matrix consisting of the eigenvalues of A.

A direct result of this theorem is

Corollary 5.1

If A is a symmetric matrix, then there exist n eigenvectors of A that form an orthonormal set.

Definition 5.3

Two $n \times n$ matrices A and B are said to be *similar* if there exists a non-singular matrix S with

$$S^{-1}AS = B$$

Theorem 5.4

Let A and B be similar matrices and λ be an eigenvalue of A with corresponding eigenvector $\mathbf{x}$. Then λ is also an eigenvalue of B and, if $S^{-1}AS=B$, then $S^{-1}\mathbf{x}$ is an eigenvector of B associated with λ.

It is often useful for us to find a bound for the magnitude of the eigenvalues of a matrix. In Chapter 4 the ability to do so would have allowed us to estimate the condition number of a matrix and, because most methods of determining eigenvalues are iterative, could be useful here in providing initial estimates of eigenvalues. The following two theorems give some indication of how this can be done.

Theorem 4.11

If λ is any eigenvalue of A, then

$$|\lambda| \leq \|A\|$$

for any matrix norm.

This theorem was proved in Chapter 4.

A more precise localisation of the eigenvalues can be obtained by using the results of the next theorem.

Theorem 5.5 (Gershgorin's circle theorem)

Let A be an $n\times n$ matrix and let C_i be the disc with centre a_{ii} and radius

$$r_i = \sum_{\substack{j=1\\ j\neq i}}^{n} |a_{ij}|$$

i.e. C_i consists of all points z such that

$$|z-a_{ii}| \leq \sum_{\substack{j=1\\ j\neq i}}^{n} |a_{ij}|$$

Let D be the union of all the discs C_i $\{i=1,2,\ldots,n\}$; then all the eigenvalues of A lie within D.

A proof of this result can be found in Burden *et al.* (1981, p.404).

5.2 Methods for a single eigenvalue

It is frequently the case that we require only a single eigenvalue or eigenvector of the matrix A; for example, in the section on the convergence of iterative methods in Chapter 4 we were interested in the size of the eigenvalue of maximum modulus. The two techniques described in this section are designed to find a single eigenvalue and the first method, the *power* method, calculates the eigenvalue of maximum modulus (the *dominant* eigenvalue) and its corresponding eigenvector.

We shall consider only the case where the dominant eigenvalue is simple (i.e. not multiple) since, otherwise, the method may not converge. See Gourlay and Watson (1973, p.40) for a discussion of this difficulty.

5.2.1 The power method

In this section it is assumed that A has a set of linearly independent eigenvectors $\boldsymbol{x}_1,\boldsymbol{x}_2,\ldots,\boldsymbol{x}_n$ corresponding to the eigenvalues $\lambda_1,\lambda_2,\ldots,\lambda_n$ and each normalised so that $\|\boldsymbol{x}_i\|_\infty=1$. It is assumed that the eigenvalues are ordered by size and that the dominant eigenvalue is distinct so that

$$|\lambda_1| > |\lambda_2| \geq \cdots \geq |\lambda_n|$$

The power method is an iterative technique for finding the dominant eigenvalue, λ_1. An initial approximation, $\boldsymbol{v}^{(0)}$, is made to the eigenvector $\boldsymbol{x}_1$ and is usually normalised so that $\|\boldsymbol{v}^{(0)}\|_\infty=1$. Now, due to the linear independence of the eigenvectors, there exist constants $\alpha_1,\alpha_2,\ldots,\alpha_n$, not all zero, such that $\boldsymbol{v}^{(0)}$ can be written as

$$\begin{aligned}\boldsymbol{v}^{(0)} &= \alpha_1\boldsymbol{x}_1 + \alpha_2\boldsymbol{x}_2 + \cdots + \alpha_n\boldsymbol{x}_n \\ &= \sum_{i=1}^{n} \alpha_i\boldsymbol{x}_i\end{aligned}$$

The iteration takes the form

$$\begin{aligned}\boldsymbol{y}^{(1)} &= A\boldsymbol{v}^{(0)} \\ &= \alpha_1 A\boldsymbol{x}_1 + \alpha_2 A\boldsymbol{x}_2 + \cdots + \alpha_n A\boldsymbol{x}_n \\ &= \alpha_1\lambda_1\boldsymbol{x}_1 + \alpha_2\lambda_2\boldsymbol{x}_2 + \cdots + \alpha_n\lambda_n\boldsymbol{x}_n\end{aligned}$$

The vector $\boldsymbol{y}^{(1)}$ can now be divided by a constant μ_1 (the element of largest modulus of $\boldsymbol{y}^{(1)}$) so that $\|\boldsymbol{y}^{(1)}/\mu_1\|_\infty = 1$ and the resulting vector, $\boldsymbol{v}^{(1)}$, satisfies

$$\boldsymbol{v}^{(1)} = \boldsymbol{y}^{(1)}/\mu_1 = \left(\sum_{i=1}^{n} \alpha_i\lambda_i\boldsymbol{x}_i\right)/\mu_1$$

Similarly

$$\boldsymbol{y}^{(2)} = A\boldsymbol{v}^{(1)} = \left(\sum_{i=1}^{n} \alpha_i\lambda_i A\boldsymbol{x}_i\right)/\mu_1$$

$$= \left(\sum_{i=1}^{n} \alpha_i \lambda_i^2 \boldsymbol{x}_i \right) / \mu_1$$

and so, normalising $\boldsymbol{y}^{(2)}$ by dividing by μ_2.

$$\boldsymbol{v}^{(2)} = \boldsymbol{y}^{(2)}/\mu_2 = \left(\sum_{i=1}^{n} \alpha_i \lambda_i^2 \boldsymbol{x}_i \right) / (\mu_1 \mu_2)$$

and in general

$$\boldsymbol{v}^{(k)} = \left(\sum_{i=1}^{n} \alpha_i \lambda_i^k \boldsymbol{x}_i \right) / (\mu_1 \mu_2 . \, . \, . \mu_k)$$

$$= \frac{\lambda_1^k \left(\alpha_1 \boldsymbol{x}_1 + \alpha_2 \left(\dfrac{\lambda_2}{\lambda_1} \right)^k \boldsymbol{x}_2 + \cdots + \alpha_n \left(\dfrac{\lambda_n}{\lambda_1} \right)^k \boldsymbol{x}_n \right)}{\mu_1 \mu_2 \cdots \mu_k} \tag{5.3}$$

Now since $| \lambda_i | / | \lambda_1 | < 1 \; \{i=2,3,. \, . \, .,n\}$ we have the result that, as k tends to infinity

$$\boldsymbol{v}^{(k)} \rightarrow \frac{\alpha_1 \lambda_1^k}{\mu_1 \mu_2 . \, . \, . \mu_k} \boldsymbol{x}_1$$

We have insisted that both $\boldsymbol{v}^{(k)}$ and $\boldsymbol{x}_1$ are normalised so that their largest element is 1 and as a result we can conclude that

$$\boldsymbol{v}^{(k)} \rightarrow \boldsymbol{x}_1 \quad \text{as} \quad k \rightarrow \infty$$

Further

$$\frac{\alpha_1 \lambda_1^k}{\mu_1 \mu_2 . \, . \, . \mu_k} \rightarrow 1 \quad \text{as} \quad k \rightarrow \infty$$

and this can be rewritten as

$$\mu_1 \mu_2 . \, . \, . \mu_k \rightarrow \alpha_1 \lambda_1^k \quad \text{as} \quad k \rightarrow \infty \tag{5.4}$$

Similarly

$$\mu_1\mu_2 \ldots \mu_k\mu_{k+1} \to \alpha_1\lambda_1^{k+1} \quad \text{as} \quad k \to \infty \tag{5.5}$$

and dividing (5.5) by (5.4) gives the result that

$$\mu_{k+1} \to \lambda_1 \quad \text{as} \quad k \to \infty$$

Unless some information about the eigenvector $\boldsymbol{x}_1$ is available an initial approximation such as

$$\boldsymbol{v}^{(0)} = (1,1,\ldots,1)^T$$

is usually sufficient to ensure convergence. It is as well, however, to be aware of a possible pitfall. If the initial approximation, $\boldsymbol{v}^{(0)}$, is *deficient* in $\boldsymbol{x}_1$, i.e. $\alpha_1=0$, then the result in (5.5) no longer holds and, if $\alpha_2 \neq 0$, successive values μ_k tend to λ_2 instead. An example illustrating this property is given in the exercises at the end of this chapter. However, rounding errors occurring during the calculation can come to the rescue; after many iterations of the power method, rounding errors will introduce a small multiple of $\boldsymbol{x}_1$ into the approximation and then subsequent approximations will converge to the dominant eigenvector.

The procedure of Fig. 5.1 implements the power iteration using $\boldsymbol{v}^{(0)}=(1,1,\ldots,1)^T$. The data types assumed are as used in Chapter 4.

```
procedure Power
    (a : nbyn;   maxits : posints;   tol : real;
     var success : boolean;   var x : vectors;
     var lambda : real;   var noofits : posints);

  var
    i : subs;
    m, max, newv : real;
    y, v : vectors;
    itcount : 0 . . maxint;
    allwithintol : boolean;

begin
  for i := 1 to n do
    v[i] := 1;
  itcount := 0;
  repeat
    itcount := itcount + 1;   max := 0;
      {Form y = Av and record maximum}
    for i := 1 to n do
```

```
        begin
          m := dotprod (a[i], v);   y[i] := m;
          if abs(m) > abs(max) then
            max := m
        end;
          { Normalise y and test for convergence }
        allwithintol := true;
        for i := 1 to n do
        begin
          newv := y[i]/max;
          if abs (newv - v[i]) > tol then
            allwithintol := false;
          v[i] := newv
        end
      until allwithintol or (itcount = maxits);

      success := allwithintol;
      if success then
      begin
        x := v;   lambda := max;   noofits := itcount
      end
    end { Power };
```

Figure 5.1

If we consider (5.3) we can see that the difference of $\boldsymbol{v}^{(k)}$ and $\boldsymbol{x}_1$ (and hence of μ_k and λ_1) is governed by the factor $|\lambda_2/\lambda_1|^k$ since λ_2 is the next largest eigenvalue; similarly the difference between $\boldsymbol{v}^{k+1}$ and x_1 is governed by $|\lambda_2/\lambda_1|^{k+1}$ so that

$$\lim_{k\to\infty} \frac{\| \boldsymbol{v}^{(k+1)} - \boldsymbol{x}_1 \|_\infty}{\| \boldsymbol{v}^{(k)} - \boldsymbol{x}_1 \|_\infty} = \frac{|\lambda_2|}{|\lambda_1|} < 1$$

We have outlined the proof of the following theorem.

Theorem 5.6

Let the eigenvalues λ_i $\{i=1,2,\ldots,n\}$ of the matrix A be ordered such that

$$|\lambda_1| > |\lambda_2| \geq |\lambda_3| \geq \cdots \geq |\lambda_n|$$

Let the initial estimate $\boldsymbol{v}^{(0)}$ of the eigenvector $\boldsymbol{x}_1$ corresponding to λ_1 not be deficient in $\boldsymbol{x}_1$. Then the iterates produced using the power method converge linearly (that is, with order 1) to the dominant eigenvalue λ_1.

This means that we can use Aitken's Δ^2 acceleration technique of Chapter 3 to speed up the convergence. We would apply the acceleration to each component of the approximations $\boldsymbol{v}^{(k)}$.

There is an alternative acceleration technique, which will give accelerated approximations to the eigenvalue alone. In terms of the power iteration the *Rayleigh quotient*, $q^{(k+1)}$, is given by

$$q^{(k+1)} = \frac{\boldsymbol{y}^{(k+1)\mathrm{T}}\boldsymbol{v}^{(k)}}{\boldsymbol{v}^{(k)\mathrm{T}}\boldsymbol{v}^{(k)}}$$

It is possible to show that $q^{(k+1)}$ tends to λ_1 but in this case with order 2, enabling the iteration to be terminated even though the eigenvector has not been calculated to the same accuracy as the eigenvalue. The modification of the procedure of Fig. 5.1 to involve the Aitken or Rayleigh acceleration is left as an exercise for the reader.

5.2.2 Deflation

Having calculated good approximations to λ_1 and its eigenvector $\boldsymbol{x}_1$ we may be interested in calculating further eigenvalues and eigenvectors. The power method can be used to do this by *deflating* the matrix A; the eigenvalue λ_1 and eigenvector $\boldsymbol{x}_1$ are 'subtracted' from A giving a matrix B with eigenvalues $\lambda_2,\lambda_3,...,\lambda_n$ and eigenvectors which are related to those of A. The following theorem illustrates one technique for achieving this.

Theorem 5.7

Suppose that z_1 is any vector with the property that $z_1^{\mathrm{T}}\boldsymbol{x}_1=1$; then the matrix

$$B = A - \lambda_1 \boldsymbol{x}_1 z_1^{\mathrm{T}}$$

has eigenvalues $0,\lambda_2,\lambda_3,...,\lambda_n$ and associated eigenvectors $\boldsymbol{x}_1,\boldsymbol{w}_2,...,\boldsymbol{w}_n$ where $\boldsymbol{x}_i$ and $\boldsymbol{w}_i$ are related by

$$\boldsymbol{x}_i = (\lambda_i - \lambda_1)\boldsymbol{w}_i + \lambda_1 (z_1^{\mathrm{T}}\boldsymbol{w}_i)\boldsymbol{x}_1 \qquad \text{for } i = 2,3,...,n$$

The proof of this result is beyond the scope of this book but may be found in Wilkinson (1965, p.596), The technique known as *Wielandt's deflation* is based on the result of this theorem and defines the vector z_1 as

$$z_1 = (a_{r1},a_{r2},...,a_{rn})^{\mathrm{T}}/(\lambda_1 x_{1r}) \qquad (x_{1r} \neq 0)$$

a multiple of some row r of A. The verification that this choice of z_1 satisfies the conditions of the theorem is left as an exercise for the reader. This choice has the great advantage that the rth row of the matrix

$$B = A - \lambda_1 \boldsymbol{x}_1 \boldsymbol{z}_1^{\mathrm{T}}$$

consists entirely of zeros. Now if λ is a non-zero eigenvalue of B associated with the eigenvector $\boldsymbol{w}$ such that

$$B\boldsymbol{w} = \lambda \boldsymbol{w}$$

then the rth element of $\boldsymbol{w}$ must be zero and so the rth column of B makes no contribution to the product $B\boldsymbol{w}$. As a result, B can be replaced by the $(n-1)\times(n-1)$ matrix B^* obtained from B by deleting the rth row and column. B^* will have eigenvalues $\lambda_2,\lambda_3,\ldots,\lambda_n$ and the power method can be applied to B^* to determine λ_2.

```
procedure Wielandt (var a : nbyn;   n, k : subs;   x : vectors);

  var
    i, j : subs;
    ak : vectors;

  procedure TransformRow (var ai : vectors;   k : subs;   xi : real);
    var
      j : subs;
  begin
    for j := 1 to n do
      if j<>k then
        ai[j] := ai[j] − xi * ak[j]
  end { Transform row };

begin {Wielandt }
  ak := a[k];
  for i := 1 to n do
    if i <> k then
      TransformRow (a[i], k, x[i]);
  for i := k to n − 1 do
    a[i] := a[i + 1];
  for i := 1 to n−1 do
    for j := k to n−1 do
      a[i,j] := a[i,j + 1]
end { Wielandt };
```

Figure 5.2

Wielandt deflation has been defined in terms of the dominant eigenvalue and its eigenvector but the technique is general in that it could be used to deflate the matrix given *any* eigenvalue–eigenvector pair. The procedure of Fig. 5.2 produces the matrix B^* given the eigenvector $\boldsymbol{x}$ corresponding to an eigenvalue λ of A. To simplify the computation, an r is chosen for which $x_r=1$, the maximum element of the eigenvector. This means that the corresponding $\boldsymbol{z}$ is given by

$$z = (a_{r1}, a_{r2}, \ldots, a_{rn})^{\mathrm{T}}/\lambda$$

and each element of B (ignoring row r and column r) is given by

$$\begin{aligned} b_{ij} &= a_{ij} - \lambda x_i z_j \\ &= a_{ij} - \lambda x_i a_{rj}/\lambda \\ &= a_{ij} - x_i a_{rj} \qquad (i,j \neq r) \end{aligned}$$

and so B^* can be computed without knowledge of λ.

In theory, this procedure could be used to find all the eigenvalues and eigenvectors of a matrix A by forming successively deflated matrices of orders $(n-1), (n-2), \ldots, 2$; the eigenvalues of a 2×2 matrix can be found directly by evaluating the characteristic polynomial and determining its roots. However, this method is prone to the accumulation of rounding errors and, when all the eigenvalues and eigenvectors are to be found, methods which make use of similarity transformations are to be preferred. These will be discussed in Section 5.3.

5.2.3 Inverse power method

On occasion, we shall be interested in finding the eigenvalue of smallest modulus. Provided that A is nonsingular, the power method can be used to determine this value by noting that the eigenvalues of the inverse of A are the reciprocals of the eigenvalues of A with exactly the same eigenvectors. For example, if $\boldsymbol{x}_i$ is an eigenvector of A corresponding to the eigenvalue λ_i, then

$$A\boldsymbol{x}_i = \lambda_i \boldsymbol{x}_i$$

Now multiplying both sides of A^{-1} gives the result

$$A^{-1}A\boldsymbol{x}_i = \boldsymbol{x}_i = \lambda_i A^{-1}\boldsymbol{x}_i$$

or

$$A^{-1}\boldsymbol{x}_i = (1/\lambda_i)\boldsymbol{x}_i \tag{5.6}$$

The choice of index i is arbitrary and so the result holds for all the eigenvalues. From (5.6) we can see that the dominant eigenvalue of A^{-1} is the reciprocal of the smallest eigenvalue of A and so, by applying the power method to A^{-1}, the reciprocal of the value we want can be found.

The *inverse power* method enables us to find the eigenvector corresponding to an eigenvalue close to a value s, without explicitly computing a matrix inverse (s may be an approximation to an eigenvalue obtained from one or two iterations of the power method.) This technique makes use of the fact that the eigenvalue of smallest modulus of the matrix

$$A - sI$$

is $\lambda - s$, where λ is the eigenvalue of A closest to s. Thus the dominant eigenvalue of $(A-sI)^{-1}$ is the reciprocal of the smallest eigenvalue of $(A-sI)$ with the same eigenvector. The inverse power iteration is

$$y^{(k+1)} = (A-sI)^{-1}v^k \tag{5.7}$$

and

$$v^{(k+1)} = y^{(k+1)}/\mu_{k+1}$$

with μ_{k+1} chosen so that $v^{(k+1)}$ has an L_∞-norm of 1.

In the process of using (5.7) we do not actually form the inverse matrix $(A-sI)^{-1}$. Instead, we solve

$$(A-sI)y^{(1)} = v^{(0)}$$

taking $v^{(0)} = (1,1,\ldots,1)^T$ for example and using LU decomposition with pivoting if necessary. We store the L and U factors as described in Chapter 4 and then, in subsequent iterations, the iterate $y^{(k+1)}$ is found by solving

$$LUy^{(k+1)} = v^{(k)}$$

The number of operations (multiplications and divisions) this involves is of order n^2 rather than of order n^3 as necessary when using, for example, Gaussian elimination. When two successive iterates $v^{(k+1)}$ and $v^{(k)}$ agree to within some specified tolerance, the process is terminated. The value μ_{k+1} is the smallest eigenvalue of $(A-sI)^{-1}$ and so the eigenvalue of A which is closest to s is

$$s + 1/\mu_{k+1}$$

and its corresponding eigenvector is $v^{(k+1)}$. This process converges rapidly if s is close to one of the eigenvalues.

The procedure of Fig. 5.3 illustrates this method. It makes use of the procedures *RecordGauss* of Fig. 4.4 and *LUsolve* of Fig. 4.7 and so does not incorporate pivoting. The data type *rowvectors* is as defined in Section 4.1.1.

```
procedure InversePower
    (a : rowvectors;   s : real;   maxits : posints;
     tol : real;   var success : boolean;
     var x : vectors;   var lambda : real;   var noofits : posints);

  var
    y : vectors;
    i : subs;
    allwithintol : boolean;
    newv, mu : real;
    itcount : posints;
    ok : boolean;
    state : (iterating, tolachieved, maxitsreached);

begin
  { Create (A -sI)y = 1 }
  for i := 1 to n do
  with a[i] do
  begin
    coeff[i] := coeff[i] - s;   b := 1
  end;

  RecordGauss (a, y, ok);

  if not ok then success := false else
  begin
    itcount := 1;   state := iterating;
    repeat
      { Determine the maximum modulus element of y }
      mu := 0;
      for i := 1 to n do
        if abs(y[i])>abs(mu) then
          mu := y[i];
      { Normalise y and test for convergence }
      allwithintol := true;
      for i := 1 to n do
      with a[i] do
      begin
      newv := y[i]/mu;
        if abs(newv-b)>tol then
          allwithintol := false;
        b := newv
      end;
      if allwithintol then state := tolachieved else
        if itcount = maxits then state := maxitsreached else
```

```
            begin
              itcount := itcount + 1;   LUsolve (a,y)
            end
        until state <> iterating;

        success := allwithintol;
        if success then
        begin
          lambda := s + 1/mu;   noofits := itcount;
          for i := 1 to n do
            x[i] := a[i].b
        end
      end
    end { Inverse power };
```

Figure 5.3

Table 5.1

i	μ_i	$v^{(i)}$			
1	0.500000	0.000000	0.000000	1.000000	1.000000
2	0.833333	−0.400000	−0.400000	1.000000	1.000000
3	0.966667	−0.482759	−0.482759	1.000000	1.000000
4	0.994253	−0.497110	−0.497110	1.000000	1.000000
5	0.999037	−0.499518	−0.499518	1.000000	1.000000
6	0.999839	−0.499920	−0.499920	1.000000	1.000000
7	0.999973	−0.499987	−0.499987	1.000000	1.000000
8	0.999996	−0.499998	−0.499998	1.000000	1.000000
9	0.999999	−0.500000	−0.500000	1.000000	1.000000
10	1.000000	−0.500000	−0.500000	1.000000	1.000000

The results given in Table 5.1 illustrate the progress of the above procedure when applied to the matrix

$$A = \begin{bmatrix} 5 & 4 & 1 & 1 \\ 4 & 5 & 1 & 1 \\ 1 & 1 & 4 & 2 \\ 1 & 1 & 2 & 4 \end{bmatrix}$$

and using a tolerance of 5E−7 to find the eigenvector corresponding to the eigenvalue closest to 4. After 10 iterations, the process converges to within this tolerance and $\mu_{10}=1$, giving the required eigenvalue as $4+1/1$ which is 5. The corresponding eigenvector is

$$(-0.500000, -0.500000, 1.000000, 1.000000)^{\mathrm{T}}$$

The inverse power method is an extremely efficient way of determining the eigenvector; this is especially true when the matrix A is tridiagonal, since the LU decomposition then requires only of order n arithmetic operations to form the factors. When the methods in the next section have been described we shall see that this is an important consideration.

5.3 Methods for the complete eigensystem

In this section we describe some methods for finding *all* the eigenvalues of the matrix A. It is possible to use these methods to evaluate eigenvectors as well but we shall concentrate on the eigenvalues. The methods are based on the result of Theorem 5.3 which states that, for a real symmetric matrix A, there exists an orthogonal matrix P such that

$$P^{-1}AP = D$$

Of course, since P is orthogonal, we could write instead

$$P^{\mathrm{T}}AP = D$$

Now, by Definition 5.3, A and D are similar and so, by Theorem 5.4, they have the same eigenvalues. Since D is a diagonal matrix, it follows that the eigenvalues of A are simply the diagonal elements of D.

Two of the methods of this section make use of *plane rotation* matrices $R(p,q)$ which are unit matrices except for the following elements

$$\begin{aligned} r_{pp} &= \cos\theta & r_{pq} &= \sin\theta \\ r_{qp} &= -\sin\theta & r_{qq} &= \cos\theta \end{aligned} \tag{5.8}$$

where the choice of θ is discussed later. $R(p,q)$ is referred to as a rotation in the plane (p,q) and it is a simple exercise to show that $R(p,q)$ is an orthogonal matrix and, as a result, R^{-1} is R^{T}. Premultiplication (postmultiplication) of A by R affects only rows (columns) p and q of A. Hence the only elements of A modified by the similarity transformation $A^*=R^{\mathrm{T}}AR$ lie in the pth and qth rows and columns. The following matrix shows the effect of a transformation, using $R(2,4)$, applied to a 5×5 matrix A

$$\begin{bmatrix} a & s & a & s & a \\ r & t & r & t & r \\ a & s & a & s & a \\ r & t & r & t & r \\ a & s & a & s & a \end{bmatrix}$$

Those elements marked a represent components of A unchanged by the transformation, those marked r have been modified only in the premultiplication by $R^T(2,4)$, those marked s have been modified only in the postmultiplication by $R(2,4)$ and those marked t have been modified in both the premultiplication and the postmultiplication. The modified elements are

$$\begin{aligned} a^*_{pp} &= a_{pp}C^2 + 2a_{pq}CS + a_{qq}S^2 \\ a^*_{qq} &= a_{pp}S^2 - 2a_{pq}CS + a_{qq}C^2 \\ a^*_{pq} &= (a_{qq}-a_{pp})CS + a_{pq}(C^2-S^2) = a^*_{qp} \\ a^*_{pj} &= a_{pj}C + a_{qj}S = a^*_{jp} \\ a^*_{qj} &= -a_{pj}S + a_{qj}C = a^*_{jq} \end{aligned} \quad \Bigg\} \quad j \neq p,q \tag{5.9}$$

where $C=\cos\theta$ and $S=\sin\theta$.

Lemma 5.1

Let R be an orthogonal matrix and let B be defined by

$$B = R^T AR$$

then

$$\| B \|_E = \| A \|_E$$

Proof

To prove this, we need the result that the sum of the eigenvalues of a matrix A is equal to the sum of its diagonal elements (the *trace* of A); this can easily be proved by considering the characteristic polynomial and its coefficients. Now

$$\| A \|^2_E = \text{trace}(A^T A) = \text{trace}(A^2) = \sum_{i=1}^{n} \lambda_i^2$$

and, since similarity transformations leave eigenvalues unchanged, we have

$$\| B \|_E^2 = \| A \|_E^2 \blacksquare$$

As a result, if we consider a sequence of matrices $A^{(k)}$, where $A^{(k+1)}$ is obtained from $A^{(k)}$ by a similarity transformation, then the Euclidean norm of each $A^{(k)}$ is the same as that of $A^{(0)} = A$. But, since we are trying to transform A into a diagonal matrix, we are trying to reduce the size of the off-diagonal elements. Therefore, if we define

$$G^{(k)} = \sum_{i=1}^{n} \sum_{j=1}^{n} | a_{ij}^{(k)} |^2 \qquad (i \neq j)$$

and

$$H^{(k)} = \sum_{i=1}^{n} | a_{ii}^{(k)} |^2$$

and then choose the parameters p, q and θ so that

$$G^{(k+1)} \leq G^{(k)}$$

then we have

$$H^{(k+1)} \geq H^{(k)}$$

The most rapid convergence of an algorithm of the above form occurs when the parameters are chosen to maximise the difference $H^{(k+1)} - H^{(k)}$ at each stage.

5.3.1 Jacobi's method

Jacobi's method is an iterative technique which uses the orthogonal plane rotations of (5.8) and reduces A to diagonal form. It can be shown that the choice of parameters in this method maximises the rate of convergence. Defining $A^{(0)} = A$, we construct the sequence of matrices

$$A^{(k+1)} = R^{\mathrm{T}}(p,q)A^{(k)}R(p,q)$$

for $k = 0,1,2,\ldots$. The choice of the plane (p,q) and the angle θ is straightforward. We search the upper triangular part of $A^{(k)}$ to find the element of maximum modulus, $a_{pq}^{(k)}$, and then choose θ to ensure that $a_{pq}^{(k+1)}$ is identically zero. By symmetry only the upper triangular part of $A^{(k)}$ need be considered. From (5.9) we can see that the equality

$$(a_{qq}^{(k)}-a_{pp}^{(k)})CS + a_{pq}^{(k)}(C^2-S^2) = 0$$

must hold. This can be rewritten as

$$-(a_{pp}^{(k)}-a_{qq}^{(k)}) \sin 2\theta + 2a_{pq}^{(k)} \cos 2\theta = 0$$

from which

$$\tan 2\theta = \frac{2a_{pq}^{(k)}}{a_{pp}^{(k)}-a_{qq}^{(k)}}$$

In the case when $a_{pp}^{(k)}=a_{qq}^{(k)}$, θ is chosen to be $(\pi/4)(a_{pq}^{(k)}/ \mid a_{pq}^{(k)} \mid)$.

The process continues with a search of $A^{(k+1)}$ for its largest off-diagonal element, which is then reduced to zero by a further plane rotation. Of course, if $a_{pq}^{(k)}$ is zero then no rotation is required and, with our choice of p and q, $A^{(k)}$ is already diagonal. It is important to note that any transformation will probably change the values of elements reduced to zero by previous transformations; hence the iterative nature of the method.

By considering equations (5.9) it is easy to show that

$$G^{(k+1)} = G^{(k)} - 2\,(a_{pq}^{(k)})^2$$

and

$$H^{(k+1)} = H^{(k)} + 2(a_{pq}^{(k)})^2$$

verifying that the sequence of matrices is tending to a diagonal matrix. The reader is referred to Gourlay and Watson (1973, p.65) for a proof of the result that the sequence tends to a *fixed* diagonal matrix, provided $\mid \theta \mid \leq \pi/4$.

The implementation of Jacobi's method, as described above, is rather time consuming since, at each stage, a complete search of the upper triangular part of $A^{(k)}$ is required to find the element $a_{pq}^{(k)}$ of maximum modulus. The *cyclic* Jacobi method does not perform this search; instead, elements are annihilated in a strict order. The usual ordering is row-wise; rotations are chosen to annihilate the elements in positions (1,2), (1,3), . . ., $(1,n)$, (2,3), . . ., $(n-1,n)$. If an element is small in comparison with the sum of squares of the diagonal elements then the rotation for that element need not be performed. This variant sacrifices the theoretical rapidity of convergence of the standard Jacobi method for improved computational efficiency. The procedure of Fig. 5.4 implements the cyclic Jacobi method.

In the procedure *JacobiTransform* the angle θ (*theta*) is calculated in a straightforward manner using the standard function *arctan*, but there is a

```
procedure CyclicJacobi
    (a : nbyn;   maxits : posints;   tol : real;
     var success : boolean;
     var lambda : vectors;   var noofits : posints);
  var
    p, q : subs;
    sumsq : real;
    itcount : 0 . . maxint;
    allwithintol : boolean;


  procedure JacobiTransform
      (p, q : subs;   var a : nbyn;   var ap, aq : vectors);
    const
      pi = 3.14159265;
    var
      app, aqq, apq, apj, aqj,
        theta, c, s, cs, csq, ssq : real;
      j : subs;
  begin
    app := ap[p];   apq := ap[q];   aqq := aq[q];
    if abs(app-aqq) > assumedzero then
    begin
      theta := arctan (2 * apq/(app-aqq)) / 2;
      if theta > pi/4 then
        theta := theta - pi/2
    end else
      if apq > 0 then theta := pi/4 else theta := -pi/4;
    {theta chosen to reduce a[p,q] and a[q,p] to zero}

    c := cos (theta);   s := sin (theta);
    csq := sqr(c);   ssq := sqr(s);   cs := c * s;
    ap[p] := app * csq + 2 * apq * cs + aqq * ssq;
    aq[q] := app * ssq - 2 * apq * cs + aqq * csq;
    ap[q] := 0;   aq[p] := 0;

    for j := 1 to n do
      if not (j in [p,q]) then
      begin
        apj := ap[j] * c + aq[j] * s;
        aqj := aq[j] * c - ap[j] * s;
        ap[j] := apj;   a[j,p] := apj;
        aq[j] := aqj;   a[j,q] := aqj
      end
  end { Jacobi transform };
```

```
begin { Cyclic Jacobi }
  itcount := 0;
  repeat
    itcount := itcount + 1;   sumsq := 0;
    for p := 1 to n do
      sumsq := sumsq + sqr(a[p,p]);
    allwithintol := true;
    for p := 1 to n − 1 do
      for q := p + 1 to n do
        if abs(a[p,q]/sumsq) > tol then
        begin
          allwithintol := false;   JacobiTransform (p, q, a, a[p], a[q])
        end
  until allwithintol or (itcount=maxits);

  success := allwithintol;
  if success then
  begin
    for p := 1 to n do
      lambda[p] := a[p,p];
    noofits := itcount
  end
end { Cyclic Jacobi };
```

Figure 5.4

---

more efficient technique which does not evaluate θ explicitly. Defining $T = \tan\theta$, we have

$$\tan 2\theta = 2T/(1-T^2)$$

and, setting $d = 1/\tan 2\theta$, T can be found by solving the quadratic equation

$$T^2 + 2dT - 1 = 0$$

The root nearer zero is chosen, to force $|\theta| \leq \pi/4$. The technique described in Section 2.1 is appropriate. Using standard trigonometry, it can be verified that

$$\cos\theta = C = 1/(1+T^2)^{1/2}$$

and hence

$$\sin\theta = S = CT$$

A further variant of the basic method is the *threshold* Jacobi method. A limited search is used to find an off-diagonal element whose modulus exceeds some threshold value and then this element is annihilated and the search continued. The level of the threshold is reduced as the computation progresses until it is smaller than some prescribed tolerance for which $A^{(k+1)}$ is 'acceptably' diagonal.

Each of these variants eventually produces a diagonal matrix

$$D = R^{\mathrm{T}}AR$$

where R is the product of all the rotation matrices $R(p,q)$ and thus is orthogonal. This equation can be rewritten as

$$AR = RD$$

from which it follows that the columns of R give the eigenvectors of A. If the eigenvector corresponding to only a single value λ_j is required, then it is probably simplest to use the inverse power method rather than store all the rotations that have been used throughout the iteration. It is usually the case that only one or two iterations of the inverse power method are required to obtain an acceptable approximation to the eigenvector.

The Jacobi method has the advantage that all the eigenvalues and eigenvectors can be calculated in one process. However, as was mentioned earlier, elements which have been reduced to zero at one stage can be made non-zero at a later stage and so the iteration is a possibly infinite process; the convergence of the method can be slow, especially for large matrices.

The next two algorithms that we describe reduce A to tri-diagonal form in a known (finite) number of steps. Then the QR method, an extremely efficient iterative technique, can be used to determine the eigenvalues.

5.3.2 Givens' method

Givens' method is a modification of Jacobi's method designed to ensure that elements set to zero by one transformation do not become non-zero during a later transformation and, because A is to be reduced only to tri-diagonal form, a maximum of $(n-1)(n-2)/2$ transformations are required.

Rotations are applied in a strict order. Starting with the first column, the elements in positions (3,1) to $(n,1)$ are annihilated by rotations in planes (2,3) to $(2,n)$. The process is then repeated for columns 2, 3, . . ., $n-2$. In general, the transformation of the jth column reduces the elements $(j+2,j)$, . . ., (n,j) by rotations in the planes $(j+1,j+2)$, . . ., $(j+1,n)$. By symmetry, the upper triangular part of A, above the super-diagonal, is also reduced to zero. It is left to the reader to verify that the transformations described above do not modify zeros introduced at previous stages.

The angle θ is chosen so that a rotation in plane (p,q) reduces the element $a_{q,p-1}$ to zero. After the transformation $R^{\mathrm{T}}(p,q)A^{(k)}R(p,q)$, the element $a_{q,p-1}^{(k+1)}$ is given by

$$a_{q,p-1}^{(k+1)} = -a_{p,p-1}^{(k)}S + a_{q,p-1}^{(k)}C$$

(see equations (5.9)) and so $a_{q,p-1}^{(k+1)}$ is zero if

$$\tan\theta = a_{q,p-1}^{(k)}/a_{p,p-1}^{(k)}$$

where, again, θ is chosen so that $|\theta| \leq \pi/4$. Obviously, if $a_{q,p-1}^{(k)}$ is zero, no rotation need be performed. If the super-diagonal element $a_{q,q+1}^{(k)}$ is zero then, to improve efficiency, the matrix can be split into two blocks each of which can then be handled separately. The following matrix illustrates the splitting in a case where the (3,4) element is zero.

$$A = \begin{bmatrix} f & f & 0 & 0 & 0 & 0 & 0 \\ f & f & f & 0 & 0 & 0 & 0 \\ 0 & f & f & 0 & 0 & 0 & 0 \\ 0 & 0 & 0 & g & g & g & g \\ 0 & 0 & 0 & g & g & g & g \\ 0 & 0 & 0 & g & g & g & g \\ 0 & 0 & 0 & g & g & g & g \end{bmatrix}$$

The procedure of Fig. 5.5 performs a Givens transformation using rotations in the plane (p,q). Its form is similar to that for a Jacobi transformation but for the choice of θ and the elements reduced to zero. Because elements once reduced to zero are not referred to again, there is no need to actually set them to zero. Consequently, the statements

aq[*p*−1] := 0;   *a*[*p*−1,*q*] := 0

do not appear in the procedure. The values of $C=\cos\theta$ and $S=\sin\theta$ are calculated very simply; if we define $T=\tan\theta$ then, by using Pythagoras, $C=1/(1+T^2)^{1/2}$ and $S=TC$.

```
procedure GivensTransform (p, q : subs;   var a : nbyn;
                                   var ap, aq : vectors);
  const
     pi = 3.14159265;
  var
     app, aqq, apq, apj, aqj, twoapqcs,
        tantheta, c, csq, s, ssq, cs : real;
     j : subs;
begin
  aqj := aq[p−1];
  if abs(aqj) > assumedzero then
```

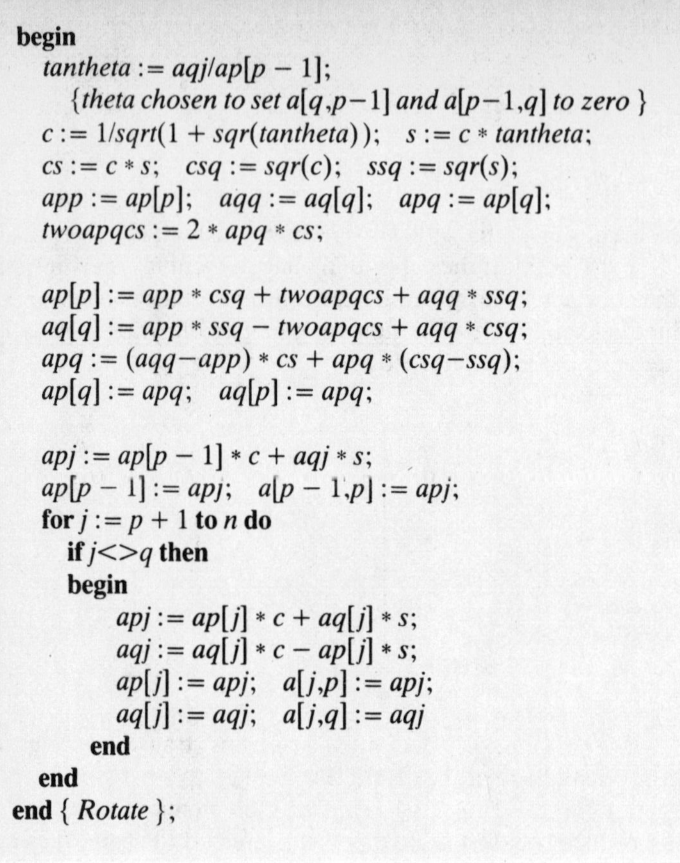

```
begin
  tantheta := aqj/ap[p - 1];
    {theta chosen to set a[q,p-1] and a[p-1,q] to zero }
  c := 1/sqrt(1 + sqr(tantheta));   s := c * tantheta;
  cs := c * s;   csq := sqr(c);   ssq := sqr(s);
  app := ap[p];   aqq := aq[q];   apq := ap[q];
  twoapqcs := 2 * apq * cs;

  ap[p] := app * csq + twoapqcs + aqq * ssq;
  aq[q] := app * ssq - twoapqcs + aqq * csq;
  apq := (aqq-app) * cs + apq * (csq-ssq);
  ap[q] := apq;   aq[p] := apq;

  apj := ap[p - 1] * c + aqj * s;
  ap[p - 1] := apj;   a[p - 1,p] := apj;
  for j := p + 1 to n do
    if j<>q then
    begin
        apj := ap[j] * c + aq[j] * s;
        aqj := aq[j] * c - ap[j] * s;
        ap[j] := apj;   a[j,p] := apj;
        aq[j] := aqj;   a[j,q] := aqj
      end
  end
end { Rotate };
```

Figure 5.5

Given this procedure, all the rotations in planes along one row can be performed using the procedure of Fig. 5.6. The whole Givens reduction to tri-diagonal form then has the form

```
for i := 2 to n - 1 do
  RowTransform (a, a[i], i)
```

This ignores the fact that the matrix may be split into blocks at some stage.

Givens' method performs approximately $4n^3/3$ multiplications in total compared with approximately $2n^3$ in general for *each* sweep of the cyclic Jacobi method. Both the Jacobi and the Givens method suffer from the disadvantage that, in order to determine the sine and cosine of the rotation, square roots need to be calculated. The Givens method requires the calculation of $(n-1)(n-2)/2$ square roots but recent work has shown that the

```
procedure RowTransform (var a : nbyn;   var ai : vectors;   i : subs);
  var
    j : subs;
begin
  for j := i + 1 to n do
    GivensTransform (i, j, a, ai, a[j])
end { Row transform };
```

Figure 5.6

method can be modified so that fewer square root calculations are necessary. For further information, see Wilkinson in Jacobs (1977, p.10).

5.3.3 Householder's method

Householder's method reduces A to tri-diagonal form by annihilating a column (and, by symmetry, a row) of elements in one operation. The technique uses elementary orthogonal matrices of the form

$$P = I - 2\boldsymbol{w}\boldsymbol{w}^{\mathrm{T}} \tag{5.10}$$

where $\boldsymbol{w}$ is a real vector with the property that

$$\boldsymbol{w}^{\mathrm{T}}\boldsymbol{w} = 1$$

It is trivial to show that $P^2=I$ and so $P^{-1}=P$ and also, from (5.10), P is symmetric with the result that $P^{\mathrm{T}}=P$. The importance of this type of matrix derives from the following result.

Lemma 5.2

Let $\boldsymbol{x}$ and $\boldsymbol{y}$ be two real vectors with

$$\boldsymbol{x}^{\mathrm{T}}\boldsymbol{x}=\boldsymbol{y}^{\mathrm{T}}\boldsymbol{y}$$

then there exists a vector $\boldsymbol{w}$ such that

$$P\boldsymbol{x}=\boldsymbol{y}.$$

Proof

The proof of this result is constructive in that we derive such a $\boldsymbol{w}$. From (5.10) we can see that

$$\boldsymbol{y} = P\boldsymbol{x} = (I-2\boldsymbol{w}\boldsymbol{w}^{\mathrm{T}})\boldsymbol{x} = \boldsymbol{x} - (2\boldsymbol{w}^{\mathrm{T}}\boldsymbol{x})\boldsymbol{w}$$

Thus $\boldsymbol{w}$ is proportional to the difference $\boldsymbol{x}-\boldsymbol{y}$, i.e.

$$\boldsymbol{w} = \frac{1}{\alpha}(\boldsymbol{x}-\boldsymbol{y}) \tag{5.11}$$

and, since $\boldsymbol{w}$ must be normalised so that $\boldsymbol{w}^{\mathrm{T}}\boldsymbol{w}=1$, α must satisfy

$$\alpha^2 = (\boldsymbol{x}-\boldsymbol{y})^{\mathrm{T}}(\boldsymbol{x}-\boldsymbol{y})$$

and so

$$\boldsymbol{w} = \frac{(\boldsymbol{x}-\boldsymbol{y})}{\{(\boldsymbol{x}-\boldsymbol{y})^{\mathrm{T}}(\boldsymbol{x}-\boldsymbol{y})\}^{1/2}}$$

The technique for reducing A to tri-diagonal form makes use of a sequence of matrices

$$P^{(k)} = I - 2\boldsymbol{w}^{(k)}\boldsymbol{w}^{(k)\mathrm{T}}$$

of the form (5.10) in which $\boldsymbol{w}^{(k)}$ has the particular form

$$\boldsymbol{w}^{(k)} = (0, \ldots, 0, w_{k+1}^{(k)}, \ldots, w_n^{(k)})^{\mathrm{T}} \tag{5.12}$$

As a result, a similarity transformation of the form $P^{(k)}AP^{(k)}$ modifies only the elements in rows $k+1$ to n and columns $k+1$ to n of A. The following matrix has been modified by a transformation involving $P^{(1)}$

$$\begin{bmatrix} a & s & s & s \\ r & t & t & t \\ r & t & t & t \\ r & t & t & t \end{bmatrix}$$

The element marked a is unchanged, those marked r have been modified only in the premultiplication by $P^{(1)}$, those marked s have been modified only in the postmultiplication by $P^{(1)}$ and those marked t have been modified in both the premultiplication and the postmultiplication. If we identify the vector $\boldsymbol{x}$ in (5.11) with the first row (or column) of A, then there will exist a matrix $P^{(1)}$ such that the third and fourth elements of the first row (and column) of $P^{(1)}AP^{(1)}$ are reduced to zero. If we define $A^{(1)} = P^{(1)}AP^{(1)}$, then let us consider also $A^{(2)} = P^{(2)}A^{(1)}P^{(2)}$ which has the form

$$\begin{bmatrix} a & a & 0 & 0 \\ a & a & s & s \\ 0 & r & t & t \\ 0 & r & t & t \end{bmatrix}$$

The premultiplication of $A^{(1)}$ by $P^{(2)}$ treats the columns (and rows) independently and so the zero elements below the sub-diagonal in the first column (and row) of $A^{(1)}$ remain zero in $A^{(2)}$. Now $P^{(2)}$ can be chosen to reduce the elements (4,2) and (2,4) to zero after which the reduced matrix will be tri-diagonal.

The choice of $\mathbf{w}^{(k)}$ in (5.12) follows directly from (5.11). Let the vector

$$\mathbf{x} = (x_1, x_2, \ldots, x_k, x_{k+1}, \ldots, x_n)^{\mathrm{T}}$$

represent the kth column (or row) of $A^{(k)}$. By this stage, the elements $x_1, x_2, \ldots, x_{k-2}$ will have been reduced to zero by previous transformations, but we shall not make use of this information. We wish to choose $\mathbf{w}^{(k)}$ so that premultiplication of $A^{(k-1)}$ by $P^{(k)}$ reduces the elements $x_{k+2}, x_{k+3}, \ldots, x_n$ to zero. If we define S^2 as

$$S^2 = x_{k+1}^2 + x_{k+2}^2 + \cdots + x_n^2$$

then the vector

$$\mathbf{y} = (x_1, x_2, \ldots, x_k, \pm S, 0, \ldots, 0)^{\mathrm{T}}$$

is such that

$$\mathbf{x}^{\mathrm{T}}\mathbf{x} = \mathbf{y}^{\mathrm{T}}\mathbf{y}$$

Hence, from (5.11),

$$\mathbf{w}^{(k)} = \frac{1}{\alpha}(\mathbf{x}-\mathbf{y}) = \frac{1}{\alpha}(0, \ldots, 0, x_{k+1} \mp S, x_{k+2}, \ldots, x_n)^{\mathrm{T}}$$

and, to reduce rounding error, the sign of S is chosen to be that of x_{k+1}. Following (5.11), the value of α is chosen so that the L_2-norm of $\mathbf{w}^{(k)}$ is unity.

$$\alpha^2 = (x_{k+1} + S)^2 + \sum_{i=k+2}^{n} x_i^2$$

$$= \sum_{i=k+1}^{n} x_i^2 + S^2 + 2x_{k+1}S = 2S^2 + 2x_{k+1}S$$

$$= 2S(S + x_{k+1})$$

For a general k, the similarity transformation has the form

$$\begin{aligned} P^{(k)}A^{(k+1)}P^{(k)} &= (I-2w^{(k)}w^{(k)\mathrm{T}})A^{(k+1)}(I-2w^{(k)}w^{(k)\mathrm{T}}) \\ &= A^{(k+1)} - 2w^{(k)}w^{(k)\mathrm{T}}A^{(k+1)} - 2A^{(k+1)}w^{(k)}w^{(k)\mathrm{T}} \\ &\quad + 4w^{(k)}w^{(k)\mathrm{T}}A^{(k+1)}w^{(k)}w^{(k)\mathrm{T}} \end{aligned}$$

but the procedure of Fig. 5.7 implements this reduction differently. It makes use of the fact that if we define the vector $v^{(k)}$ to be

$$v^{(k)} = A^{(k-1)}w^{(k)}$$

then

$$A^{(k-1)}P^{(k)} = A^{(k-1)} - 2v^{(k)}w^{(k)\mathrm{T}}$$

If we define the vector $q^{(k)}$ by

$$q^{(k)} = v^{(k)} - (w^{(k)\mathrm{T}}v^{(k)})w^{(k)}$$

then we have that

$$P^{(k)}A^{(k-1)}P^{(k)} = A^{(k-1)} - 2w^{(k)}q^{(k)\mathrm{T}} - 2q^{(k)}w^{(k)\mathrm{T}}$$

and this gives the most efficient implementation of the Householder transformations.

The steps performed by the procedure at each stage k are as follows:

1. Form w (from elements $k+1$ to n of row k of A) and update $a_{k,k+1}$ and $a_{k+1,k}$.
2. Form $v = Aw$ (only elements $k + 1$ to n are non-zero).
3. Form $u = w^{\mathrm{T}}v$ (only elements $k + 1$ to n are non-zero).
4. Form $q = v - uw$ (only elements $k + 1$ to n are non-zero; q overwrites v).
5. Form $A = A - 2wq^{\mathrm{T}} - 2qw^{\mathrm{T}}$.

As for Givens, the elements reduced to zero at each stage are not referred to again and so need not actually be set to zero. Consequently, at stage k, the procedure assigns no values to elements

$$\begin{matrix} a_{k,k+2} \; a_{k,k+3} \cdots a_{kn} \\ a_{k+2,k} \; a_{k+3,k} \cdots a_{nk} \end{matrix}$$

```
procedure Householder (var a : nbyn);
  var
    i, k : subs;
    u : real;
    w, v : vectors;
  function dotprod (u, v : vectors; m : subs) : real;
    var
      i : subs;
      dot : real;
  begin
    dot := 0;
    for i := m to n do
      dot := dot + u[i] * v[i];
    dotprod := dot
  end { dot prod };
  procedure Formw (var w, ak : vectors;
      var akplus1k : real;   kplus1 : subs);
  {Forms w_{k+1},. . .,w_n and updates a_{k,k+1} and a_{k+1,k}}
    var
      j : subs;
      s, alpha, sumsq : real;
  begin
    sumsq := 0;
    for j := kplus1 to n do
      sumsq := sumsq + sqr(ak[j]);
    if akplus1k < 0 then s := - sqrt(sumsq)
                    else s := + sqrt(sumsq);
    alpha := sqrt (2 * s * (s + akplus1k));
    w[kplus1] := (akplus1k + s) / alpha;
    for j := kplus1 + 1 to n do
      w[j] := ak[j]/alpha;
    ak[kplus1] := -s;   akplus1k := -s
  end { Form w };
  procedure Rowofa
      (var a : nbyn;   var ai : vectors;   i : subs;   twovi, twowi : real);
    var
      j : subs;
  begin
    ai[i] := ai[i] - twovi * twowi;
    for j := i+1 to n do
    begin
      ai[j] := ai[j] - v[j] * twowi - twovi * w[j];
      a[j,i] := ai[j]
    end
  end {Row of a}
```

```
begin { Householder }
  for k := 1 to n−2 do
  begin
    Formw (w, a[k], a[k + 1,k], k + 1);
    for i := k + 1 to n do
      v[i] := dotprod (a[i],w,k + 1);
    u := dotprod (w,v,k + 1);
    for i := k + 1 to n do
      v[i] := v[i] − u * w[i];
    for i := k + 1 to n do
      Rowofa (a, a[i], i, 2 * v[i], 2 * w[i])
  end
end { Householder };
```

Figure 5.7

For a more detailed discussion of the organisation of the method refer to Wilkinson and Reinsch (1971).

Householder's method requires the calculation of only one square root per column (to form S), $n-2$ square roots in all, and requires only half the multiplications of Givens' method and so is preferred, in general, whenever the matrix is dense (most of its elements non-zero). Givens' method is still useful whenever there are zero elements in the matrix because then no reduction of those elements is necessary, but Householder's method would be unable to take account of this. In addition, recent advances in the implementation of Givens' method, mentioned in Section 5.3.2, have made it as efficient as that of Householder.

5.3.4 The QR method

The QR algorithm finds all the eigenvalues of a matrix and is the most generally recommended method when the matrix is symmetric and tridiagonal. Before we describe the algorithm, we digress and reconsider matrix factorisations.

Definition 5.4

A matrix Q is *unitary* if $\bar{Q}^T Q = I$ where $\bar{Q}$ is the complex conjugate of Q.

The following theorem is fundamental to the development of the QR method.

Theorem 5.8

An arbitrary matrix A can be reduced to upper triangular form by means of a premultiplication by a unitary matrix Q.

A proof of this result can be found in Ralston and Rabinowitz (1978, p.522).

In our case the matrix A is real and symmetric and so Q is real and need only be orthogonal. Thus

$$A = QR$$

where R is upper triangular. For historical reasons the notation R is used rather than U; Francis, the originator of the method, thought of the triangular matrices as being left (L) and right (R) rather than lower (L) and upper (U). This factorisation is unique if the diagonal elements are made non-negative (the orthogonal transformations are unique up to a sign).

The usual method for obtaining such a QR factorisation is the Householder transform and so Q is the product of several Householder matrices $P^{(k)}$ of the form (5.10). In our case, however, A is real, symmetric and tri-diagonal and so only the sub-diagonal has to be reduced to zero in order to transform A to triangular form. Thus the most efficient method of performing the reduction is to use plane rotations in the planes (1,2), (2,3),. . .,($n-1$,n) to reduce the elements in positions (2,1), (3,2),. . .,(n,$n-1$) to zero. Since only the sub-diagonal is being annihilated, and not the super-diagonal, no post-multiplication of A is involved. The elements modified by the premultiplication by $R^{\mathrm{T}}(p,q)$ are

$$\begin{aligned} a^*_{pj} &= a_{pj}C - a_{qj}S \\ a^*_{qj} &= a_{pj}S + a_{qj}C \end{aligned}$$

The element a^*_{qp} is given by $a_{pp}S+a_{qp}C$ and so is zero if

$$\tan\theta = -a_{qp}/a_{pp}$$

Now we return to the problem of determining the eigenvalues of our tri-diagonal matrix. If, as before, we define the matrix $A^{(0)}$ to be equal to A then, at stage k of the process, $A^{(k-1)}$ is decomposed to give

$$A^{(k-1)} = Q^{(k-1)}R^{(k-1)} \tag{5.13}$$

and then $A^{(k)}$ is defined by

$$A^{(k)} = R^{(k-1)}Q^{(k-1)} \tag{5.14}$$

The decomposition in (5.13) is actually determined by finding the orthogonal matrix $\{Q^{(k-1)}\}^{\mathrm{T}}$ which satisfies

$$\{Q^{(k-1)}\}^{\mathrm{T}}A^{(k-1)} = R^{(k-1)}$$

It then follows that, in (5.14)

$$A^{(k)} = \{Q^{(k-1)}\}^{\mathrm{T}} A^{(k-1)} Q^{(k-1)}$$

and so $A^{(k)}$ is similar to $A^{(k-1)}$ (and hence to $A^{(0)}$) and has the same eigenvalues. A further important consequence of this similarity is that $A^{(k)}$ has the same form as $A^{(k-1)}$ in that, because $A^{(0)}=A$ is tri-diagonal, each $A^{(k)}$ $\{k=1,2,\ldots\}$ is also tri-diagonal. This means that every reduction (5.13) can be performed very efficiently.

The final and most important result is that, as k tends to infinity, the matrices $A^{(k)}$ tend to a diagonal matrix with the eigenvalues of A on the diagonal. The proof of this result is well beyond the scope of this book but can be found in Wilkinson (1965, p.963).

In general this method does not converge sufficiently rapidly to be useful in practice and this has provoked the development of the more rapidly converging *shifted QR* algorithm. At the kth stage, the following steps are taken

$$A^{(k-1)} - s^{(k-1)} I = Q^{(k-1)} R^{(k-1)}$$
$$A^{(k)} = s^{(k-1)} I + R^{(k-1)} Q^{(k-1)}$$

for some suitably chosen value $s^{(k-1)}$. A simple algebraic manipulation will show that $A^{(k)}$ is still similar to $A^{(k-1)}$.

As the iteration progresses, both $a_{n-1,n}^{(k-1)}$ and $a_{n,n-1}^{(k-1)}$ tend to zero and, when these values are acceptably small, we can take $a_{nn}^{(k-1)}$ to be an eigenvalue of A and then omit the last row and column of $A^{(k-1)}$. If $a_{n-2,n-1}^{(k-1)}$ becomes negligible, then two eigenvalues can be obtained directly from the 2×2 matrix at the bottom right-hand corner of $A^{(k-1)}$ and then the last two rows and columns of $A^{(k-1)}$ deleted. Thus the algorithm leads to a reduction of the order of the matrices and this results in very rapid convergence. Again, the proof of the convergence of the shifted QR algorithm can be found in Wilkinson (1965). The value of $s^{(k-1)}$ is usually chosen to be that eigenvalue of the matrix

$$\begin{bmatrix} a_{n-1,n-1}^{(k-1)} & a_{n-1,n}^{(k-1)} \\ a_{n,n-1}^{(k-1)} & a_{nn}^{(k-1)} \end{bmatrix}$$

which is closer to $a_{nn}^{(k-1)}$ because this choice generally produces the most rapid convergence.

The procedure of Fig. 5.8 implements the shifted QR algorithm. To permit slicing the matrix Q is stored in transposed form and so the procedure

MultiplyMatTrans can follow the lines of Fig. 1.14 but without the need to transpose the second matrix; the transpose is already available. This procedure takes no account of the special shape of the matrices involved and so assumes that all zeros are stored explicitly. As the computation proceeds the elements of R overwrite those of A.

```
procedure QR (a : nbyn;   tol : real;   var lambda: vectors);

  var
    b, qt : nbyn;
    i, j : subs;
    valsfound : 0 . . 2;
    am1m, amm, am1m1, s, s0 : real;
    m : 0 . . n;

  procedure Rotate (p, q : subs;   var ap, aq, rp, rq : vectors);
      {q = p + 1
       The two vectors ap and aq hold rows p and q of the
       matrix A and rp and rq hold rows p and q of Q^T}
    const
      pi = 3.14159265;
    var
      tantheta, s, c, aqp, aqj, apj, rpj : real;
      i, j : subs;
  begin
    aqp := aq[p];
    if abs(aqp) > assumedzero then
    begin
      tantheta := -aqp/ap[p];
        {theta chosen to set a[q,p] to zero}
      c := 1/sqrt(1 + sqr(tantheta));   s := c * tantheta;
      for j := 1 to n do
      begin
        apj := ap[j];   aqj := aq[j];
        ap[j] := c * apj - s * aqj;
        if j <> p then
          aq[j] := s * apj + c * aqj
      end;
      aq[p] := 0;

        {Now premultiply Q by R(p,q)}
      rp[q] := -s;   rq[q] := c;
      for j := 1 to p do
```

```
        begin
          rpj := rp[j];
          rp[j] := rpj * c;   rq[j] := rpj * s
        end
      end
    end { Rotate };

  begin { QR }
    m := n;
    repeat
      if m <= 2 then valsfound := m else
      begin
        valsfound := 0;
        repeat
              { Check bottom right-hand corner }
          if abs(a[m - 2,m - 1]) <= tol then valsfound := 2 else
            if abs(a[m - 1,m]) <= tol then valsfound := 1 else
            begin
                { Eigenvalues not yet found so iterate }
              amm := a[m,m];   am1m1 := a[m - 1, m - 1];
              QuadSolve (1, - (am1m1 + amm),
                  amm * am1m1 - sqr(a[m - 1,m]), s, s0);
              if abs(s0-amm) < abs(s-amm) then s := s0;
              for i := 1 to m do
                a[i,i] := a[i,i] - s;
                { Set up identity matrix }
              for i := 1 to m do
                for j := 1 to m do
                  qt[i,j] := ord(i=j);
              for i := 1 to m - 1 do
                Rotate (i, i + 1, a[i], a[i + 1], qt[i], qt[i + 1]);
              MultiplyMatTrans (a, qt, m, b);
              a := b;
              for i := 1 to m do
                a[i,i] := a[i,i] + s
            end
        until valsfound > 0
      end;

    amm := a[m,m];
    case valsfound of
      1 : lambda[m] := amm;
      2 : begin
            am1m := a[m - 1,m];   am1m1 := a[m - 1,m - 1];
            if abs(am1m) <= tol then
```

```
          begin
            lambda[m] := amm;   lambda[m - 1] := am1m1
          end else
            QuadSolve (1, -(am1m1 + amm),
              amm * am1m1 - sqr(am1m),
              lambda[m], lambda[m - 1])
        end
      end { case };
      m := m - valsfound
    until m = 0
  end { QR };
```

Figure 5.8

The matrix A is symmetric tri-diagonal, Q is upper Hessenberg (see Definition 5.5, p.175) and R has only the main diagonal and first two super-diagonals non-zero. The efficiency of the procedure of Fig. 5.8 can be significantly improved by avoiding reference to elements known to be zero. For example the process *MultiplyMatTrans*, which forms $A^{(k)}$ given $R^{(k-1)}$ and the transpose of $Q^{(k-1)}$, could be written as in Fig. 5.9 assuming that the transformation process has been similarly modified.

This makes no use of slicing and assumes the existence of the following work variables

```
var
  i, j, k : subs;
  sigma : real;
```

```
for i := 1 to m - 2 do
begin
  for j := i to i + 1 do
  begin
    sigma := 0;
    for k := i to j + 1 do
      sigma := sigma + a[i,k] * qt[j,k];
    b[i,j] := sigma
  end;
  b[i + 1,i] := sigma
end;
b[m-1,m-1] := a[m-1,m-1]*qt[m-1,m-1] + a[m-1,m]*qt[m-1,m];
sigma := a[m-1,m-1]*qt[m,m-1] + a[m-1,m]*qt[m,m];
b[m-1,m] := sigma;   b[m,m-1] := sigma;
b[m,m] := a[m,m]*qt[m,m]
```

Figure 5.9

Further improvements result on realising that the variable b, used to hold the product matrix and later assigned to a, is unnecessary; the variable a can be updated as multiplication proceeds. This has been done in Fig. 5.10 using the following work variables

```
var
   i : subs;
   aii, aiiplus1, qtmm : real;
```

```
for i := 1 to m - 2 do
begin
   aii := a[i,i];   aiiplus1 := a[i,i + 1];
   a[i,i] := aii * qt[i,i] + aiiplus1 * qt [i,i + 1];
   aiiplus1 := aii * qt[i + 1, i] + aiiplus1 * qt[i + 1,i + 1]
                  + a[i,i + 2] * qt[i + 1,i + 2];
   a[i,i + 1] := aiiplus1;   a[i + 1,i] := aiiplus1
end;
aii := a[m - 1,m - 1];   aiiplus1 := a[m - 1,m];
a[m - 1, m - 1] := aii * qt[m - 1,m - 1] + aiiplus1 * qt [m - 1,m];
qtmm := qt[m,m];
aiiplus1 := aii * qt[m,m - 1] + aiiplus1 * qtmm;
a[m - 1,m] := aiiplus1;   a[m,m - 1] := aiiplus1;
a[m,m] := a[m,m] * qtmm
```

Figure 5.10

Efficiency can be improved even further. Because A is symmetric there is no need to record both the sub-diagonal and the super-diagonal and, rather than store A and R as $n \times n$ matrices, both space and time can be saved by using vectors and storing only the relevant diagonals.

5.4 Non-symmetric matrices

Not surprisingly, the problem of determining the eigenvalues and eigenvectors of non-symmetric matrices is more difficult than that for symmetric matrices. For this reason we give only the briefest outline of how the methods already discussed can be applied to these matrices.

All the techniques described in Section 5.3 can easily be extended to find the eigenvalues of Hermitian matrices. (Hermitian matrices are the complex equivalent of real symmetric matrices; A is Hermitian if $\bar{A}^T = A$). The rotations $R(p,q)$ involved in Jacobi's and Givens' methods become complex of the form

$$\begin{aligned} r_{pp} &= e^{ix}\cos\theta & r_{pq} &= e^{iy}\sin\theta \\ r_{qp} &= -e^{-iy}\sin\theta & r_{qq} &= e^{-ix}\cos\theta \end{aligned}$$

and

$$r_{ij} = \delta_{ij} \qquad \text{for } i,j \neq p,q$$

The matrices involved in the Householder transformation become elementary unitary matrices which have the form

$$P^{(k)} = I - \mu w^{(k)}\{\bar{w}^{(k)}\}^{T}$$

where μ is a complex constant.

Lemma 5.3

$P^{(k)}$ is unitary if

$$\mu + \bar{\mu} = \mu\bar{\mu}$$

and Hermitian if $\mu=2$.

The proof of this lemma is left as an exercise for the reader.

If A is an arbitrary real matrix the situation is more difficult to handle. In this case the methods are, in general, based on the following theorem due to Schur (see Johnson and Riess, 1982, p.136).

Theorem 5.9

For any square matrix A there exists a unitary matrix R such that

$$\bar{R}^{T}AR = T$$

where T is a triangular matrix with the eigenvalues on the diagonal.

The QR algorithm can be used to determine the eigenvalues. Because of the lack of symmetry, Givens or Householder transformations are now used to reduce A to *upper Hessenberg* form rather than to tri-diagonal form.

Definition 5.5

An $n \times n$ matrix A is said to be *upper* (*lower*) *Hessenberg* if the a_{ij} element is zero for $i>j+1$ ($i+1<j$).

Both the Givens and Householder methods employ unitary transformations to reduce to zero those elements below the main sub-diagonal of A. The QR algorithm is applied to this upper Hessenberg matrix, usually with the restriction that the diagonal elements of R should be real and that Q is a unitary (rather than orthogonal) matrix. An ingenious modification of the QR algorithm, due to Francis (see Gourlay and Watson, 1973, Section 13.3), ensures that even complex eigenvalues can be found using real arithmetic by employing a double shift. In this case the matrix T is not strictly upper triangular but is *block* upper triangular with 1×1 or 2×2 blocks on the diagonal. The 1×1 blocks correspond to real eigenvalues and the 2×2 blocks correspond to pairs of complex conjugate eigenvalues.

5.5 Exercises

1 Prove the results of Theorems 5.1 and 5.2.

2 Use the procedure *Power* of Fig. 5.1 to determine the dominant eigenvalue of the matrix

$$\begin{bmatrix} 2 & -1 & 0 \\ -1 & 2 & -1 \\ 0 & -1 & 2 \end{bmatrix}$$

3 Modify the procedure *Power* of Fig. 5.1 to determine the dominant eigenvalue of the matrix in Exercise 2 but this time using $\mathbf{v}^{(0)}=(1.207107, 0.707107, -0.207107)^{\mathrm{T}}$ as the starting vector. The calculation should be repeated using several different values for *maxits* and *tol*.

4 Modify *Power* to incorporate either of the acceleration techniques discussed in Section 5.2.1 and apply the resulting procedure to determine the dominant eigenvalue of the matrix

$$\begin{bmatrix} 5 & 4 & 1 & 1 \\ 4 & 5 & 1 & 1 \\ 1 & 1 & 4 & 2 \\ 1 & 1 & 2 & 4 \end{bmatrix}$$

Compare the rate of convergence with that of the original power iteration.

5 Verify that Wielandt's choice of the vector $\mathbf{z}_1$ does satisfy the condition of Theorem 5.7.

6 Apply the inverse power method to the matrix of Exercise 2 to determine the eigenvector corresponding to the eigenvalue closest to 1.9.

7 Use the procedure *CyclicJacobi* of Fig. 5.4 to determine all the eigenvalues of the matrix of Exercise 4.

8 Write a procedure to use the threshhold Jacobi method to determine the eigenvalues of a matrix and use this procedure to determine the eigenvalues of the matrix of Exercise 4.

9 Verify that Givens' transformation does not modify zeros already transformed.

10 Verify that the Householder transformation can be written in the form used in the procedure of Fig. 5.7.

11 Use both of the procedures *GivensTransform* (Fig. 5.5) and *Householder* (Fig. 5.7) to tri-diagonalise the matrix

$$\begin{bmatrix} 1 & 2 & 1 & 2 \\ 2 & 2 & -1 & 1 \\ 1 & -1 & 1 & 1 \\ 2 & 1 & 1 & 1 \end{bmatrix}$$

and then use the procedure *QR* (Fig. 5.8) in each case to determine the eigenvalues of the matrix.

12 Prove the result of Lemma 5.3.

Chapter 6 DISCRETE FUNCTION APPROXIMATION

During an experiment it is often the case that readings of some kind (temperature or resistance for example) are taken at discrete (possibly equidistant) times. Thus at a set of points $x_0, x_1, \ldots, x_n$ we will have recorded the values $f_0, f_1, \ldots, f_n$ of a function $f(x)$, say. We may now be asked the question 'What do you think the value of f should be at the point $\bar{x}$?' where $\bar{x}$ is not one of our tabular points. A contrasting problem arises when we are given data values which are exact except, perhaps, for rounding or truncation errors (experimental readings automatically include some error). For example, a set of values may be taken from logarithmic tables and the above question could again be asked.

These two problems have to be treated in different ways. We shall attempt to solve the second problem by using *interpolation*, but this technique is not appropriate for the first because of the errors involved. We shall expand upon this difficulty as we describe the various techniques.

In both cases we shall use polynomials to approximate the function $f(x)$. Polynomials are popular as approximating functions because they are convenient to use on computers and because most people have some idea of how polynomials behave. There is a theoretical justification, based on the following theorem by Weierstrass (see Ralston and Rabinowitz, 1978, p.36), for using polynomials.

Theorem 6.1

If $f(x)$ is defined and continuous on the interval $[a,b]$ and a constant $\epsilon>0$ is given, then there exists a polynomial, $p(x)$, defined on $[a,b]$ with the property that

$$|f(x)-p(x)| < \epsilon \qquad \text{for all } x \text{ in } [a,b]$$

The proof of this theorem is beyond the scope of this book and, indeed, the result is of only passing interest. This theorem is not constructive; it gives no indication of the degree of the polynomial nor yet how to obtain its equation. However, the result is useful in the sense that we are not necessarily being stupid in using polynomials for approximation.

6.1 Polynomial interpolation

As was mentioned above, polynomial interpolation is used when the data is exact. The idea is extremely simple; we determine a polynomial which passes through all the data points and then approximate $f(\bar{x})$ by the value of the polynomial at the point $\bar{x}$. We have made the assumption that the discrete points $x_0, x_1, \ldots, x_n$ are ordered so that $x_0 < x_1 < \cdots < x_n$ and that the point $\bar{x}$ lies inside the interval $[x_0, x_n]$. If $\bar{x}$ lies outside this interval, the process is the same but is known as *extrapolation*. For reasons which will become apparent later, extrapolation is full of pitfalls and we shall consider only interpolation.

Theorem 6.2

Let x_i $\{i = 0,1,\ldots,n\}$ be $n+1$ distinct points and let f_i $\{i = 0,1,\ldots,n\}$ be any set of $n+1$ real numbers. Then there exists a unique polynomial

$$p_n(x) = a_n x^n + a_{n-1}x^{n-1} + \cdots + a_1 x + a_0 \tag{6.1}$$

of degree at most n such that $p_n(x_i) = f_i$ $\{i = 0,1,\ldots,n\}$.

Proof

In Section 6.1.1 we shall prove the existence of such a polynomial by constructing it. It remains for us to prove that the polynomial is unique and we do this by assuming the converse.

Let there be two polynomials $p_n(x)$ and $q_n(x)$, of degree at most n, interpolating the data and define the difference $d_n(x)$ by

$$d_n(x) = p_n(x) - q_n(x)$$

Thus $d_n(x)$ is also a polynomial of degree at most n and as such has at most n zeros. However, $d_n(x)$ is zero whenever $x = x_0, x_1, \ldots, x_n$, since at each of these points $p_n(x_i) = q_n(x_i) = f_i$. Thus we have shown that $d_n(x)$ has at least $n+1$ zeros and we have a contradiction which can be resolved only if $d_n(x)$ is identically zero everywhere. Thus

$$p_n(x) = q_n(x)$$

and the interpolation polynomial is therefore unique. ■

We now have to decide how to determine the equation of the polynomial $p_n(x)$ or, at least, its value at $x = \bar{x}$. One method would be to evaluate (6.1) at each data point and set the result to be equal to the function value at that point. Thus we would have $n+1$ linear equations in the $n+1$ unknowns $a_0, a_1, \ldots, a_n$ and these equations could be solved using Gaussian elimination, for example. However, this method of determining the polynomial is a classically ill-conditioned one. The matrix of the system of linear equations is related to the Hilbert matrix introduced in the exercises of Chapter 4 and, as we have seen, the Hilbert matrix is badly conditioned. Instead we use the following technique.

6.1.1 Lagrangian interpolation

In Lagrangian interpolation, the polynomial is written in the form

$$p_n(x) = f_0 l_0(x) + f_1 l_1(x) + \cdots + f_n l_n(x) \tag{6.2}$$

where each function $l_i(x)$ $\{i=0,1,\ldots,n\}$ is itself a polynomial of degree n and has the property that

$$l_i(x_j) = \delta_{ij} = \begin{cases} 1 \text{ if } i=j \\ 0 \text{ if } i \neq j \end{cases} \qquad i,j = 0,1,\ldots,n \tag{6.3}$$

where we recall that the x_j are the x-data points. Thus (6.2) is the equation of a polynomial of degree n and, using the result of (6.3), we have that

$$p_n(x_j) = f_j \qquad j = 0,1,\ldots,n$$

so that $p_n(x)$ *is* the interpolation polynomial. Each of the polynomials $l_i(x)$ has the form

$$l_i(x) \equiv \frac{(x-x_0)(x-x_1)\cdots(x-x_{i-1})(x-x_{i+1})\cdots(x-x_n)}{(x_i-x_0)(x_i-x_1)\cdots(x_i-x_{i-1})(x_i-x_{i+1})\cdots(x_i-x_n)}$$

in which the numerator is the product of all the factors $(x-x_j)$ excluding the term $(x-x_i)$ {i.e. the one with the same subscript as $l_i(x)$} and the denominator is simply the numerator evaluated at $x=x_i$.

We now present an example to show how the polynomial (6.2) is used. Let us consider interpolating for $f(0.14)$ given the values of $f(x)=e^x$ shown in Table 6.1.

Table 6.1

i	0	1	2	3
x_i	0	0.1	0.3	0.6
$f(x_i)$	1.00000	1.10517	1.34986	1.82212

First, we evaluate each of the polynomials $l_i(x)$ of (6.2) in turn at the point $x = 0.14$.

$$l_0(0.14) = \frac{(x-x_1)(x-x_2)(x-x_3)}{(x_0-x_1)(x_0-x_2)(x_0-x_3)} = \frac{(0.04)(-0.16)(-0.46)}{(-0.1)(-0.3)(-0.6)} = -0.16356$$

$$l_1(0.14) = \frac{(x-x_0)(x-x_2)(x-x_3)}{(x_1-x_0)(x_1-x_2)(x_1-x_3)} = 1.03040$$

$$l_2(0.14) = \frac{(x-x_0)(x-x_1)(x-x_3)}{(x_2-x_0)(x_2-x_1)(x_2-x_3)} = 0.14311$$

$$l_3(0.14) = \frac{(x-x_0)(x-x_1)(x-x_2)}{(x_3-x_0)(x_3-x_1)(x_3-x_2)} = -0.00996$$

$$\begin{aligned}\text{Now } f(0.14) &\simeq f_0 l_0(0.14) + f_1 l_1(0.14) + f_2 l_2(0.14) + f_3 l_3(0.14)\\ &= (1.00000)(-0.16356) + (1.10517)(1.03040)\\ &\quad + (1.34986)(0.14311) + (1.82212)(-0.00996)\\ &= 1.150251\end{aligned}$$

To six decimal places the correct value for $f(0.14)=e^{0.14}$ is 1.150274 and we will need to derive some method of calculating in advance how good an approximation to $f(x)$ we will obtain. Before we do this, we should notice that, if we sum the numerical values of the polynomials $l_i(0.14)$, we get

$$\begin{aligned}&l_0(0.14) \;+\; l_1(0.14) \;+\; l_2(0.14) \;+\; l_3(0.14) \;=\\ &(-0.16356) + (1.03040) + (0.14311) + (-0.00996) = 0.99999\end{aligned}$$

or, to within rounding error, 1.0000. This is a consequence of the fact that the interpolating polynomial must be able to reproduce the function $f(x)$ exactly if it is a polynomial of degree n or less; in particular, if $f(x)$ is constant, equal to c everywhere, then

$$f(x) = c = \sum_{i=0}^{n} c l_i(x)$$

that is,

$$\sum_{i=0}^{n} l_i(x) = 1$$

for *any* x. This is useful as a check during hand computation but is less necessary in a computer program. The function of Fig. 6.1 will approximate $f(\bar{x})$ given the set of values (x_i, f_i) $\{i=0,1,\ldots,n\}$ and assumes the environment

```
const
  n = . . .;
type
  subs = 0 . . n;
  vectors = array [subs] of real;
```

```
function Lagrange (xbar : real;   x, f : vectors) : real;
  var
    i, j : subs;
    lag, li, xi : real;
begin
  lag := 0;
  for i := 0 to n do
  begin
    li := 1;   xi := x[i];
    for j := 0 to n do
      if j<>i then
        li := li * (xbar−x[j])/(xi−x[j]);
    lag := lag + f[i] * li
  end;
  Lagrange := lag
end { Lagrange };
```

Figure 6.1

Theorem 6.3

The error in Lagrangian interpolation

$$E(x) = f(x) - p_n(x)$$

is given by

$$E(x) = \frac{(x-x_0)(x-x_1)\cdots(x-x_n)}{(n+1)!} f^{(n+1)}(\xi) \tag{6.4}$$

where ξ is dependent on x and lies in the range $x_0<\xi<x_n$.

A proof of this result can be found in Burden *et al.* (1981, p.83).

The form of (6.4) means that the error cannot be evaluated exactly since we do not know the precise dependence of ξ on x. However, we may be able to bound the derivative of f to give upper and lower bounds on the magnitude of $E(x)$.

$$|E(x)| \le \frac{|(x-x_0)(x-x_1)\cdots(x-x_n)|}{(n+1)!} \max_{x_0<x<x_n} |f^{(n+1)}(x)|$$

and

$$|E(x)| \ge \frac{|(x-x_0)(x-x_1)\cdots(x-x_n)|}{(n+1)!} \min_{x_0<x<x_n} |f^{(n+1)}(x)|$$

For the numerical example given above, the error formula is

$$E(x) = \frac{(x-x_0)(x-x_1)(x-x_2)(x-x_3)}{4!} f^{(4)}(\xi), \qquad x_0<\xi<x_3$$

and we know that $f(x)$ is e^x and so is each of its derivatives. Thus we can bound the error as follows:

$$|E(0.14)| \leq \frac{|(0.14-0.00)(0.14-0.10)(0.14-0.30)(0.14-0.60)|}{4!} \times \max_{0.00 \leq x \leq 0.60} |e^x|$$

$$= 0.000025$$

and

$$|E(0.14)| \geq \frac{|(0.14)(0.04)(-0.16)(-0.46)|}{24} \min_{0.00 \leq x \leq 0.60} |e^x|$$

$$= 0.000017$$

We can see that the actual error, $1.150274-1.150251=0.000023$, lies within the error bounds that we have calculated.

The error formula is also useful when we can choose the positions of the points x_i $\{i = 0,1,...,n\}$. A sensible choice of these points is one that minimises the interpolation error by making the magnitude of the product

$$(x-x_0)(x-x_1)\cdots(x-x_n)$$

as small as possible.

Theorem 6.4

The value of

$$|(x-x_0)(x-x_1)\cdots(x-x_n)|$$

is minimised by choosing the points x_i $\{i= 0,1,...,n\}$ to be the roots of the Chebyshev polynomial of degree $n+1$.

A proof of this result can be found in Burden *et al.* (1981, p.346).

By examining the error formula, we can see that if x lies outside the interval $[x_0,x_n]$ (recall that this is known as extrapolation) then the product of the terms $(x-x_i)$ is larger and the point ξ may lie outside the interval, in which case we have *no* information about the behaviour of the function and its derivatives. This means that we have no real idea about the accuracy of extrapolation and, in fact, practical experience and the error estimates do imply that it is an inaccurate process.

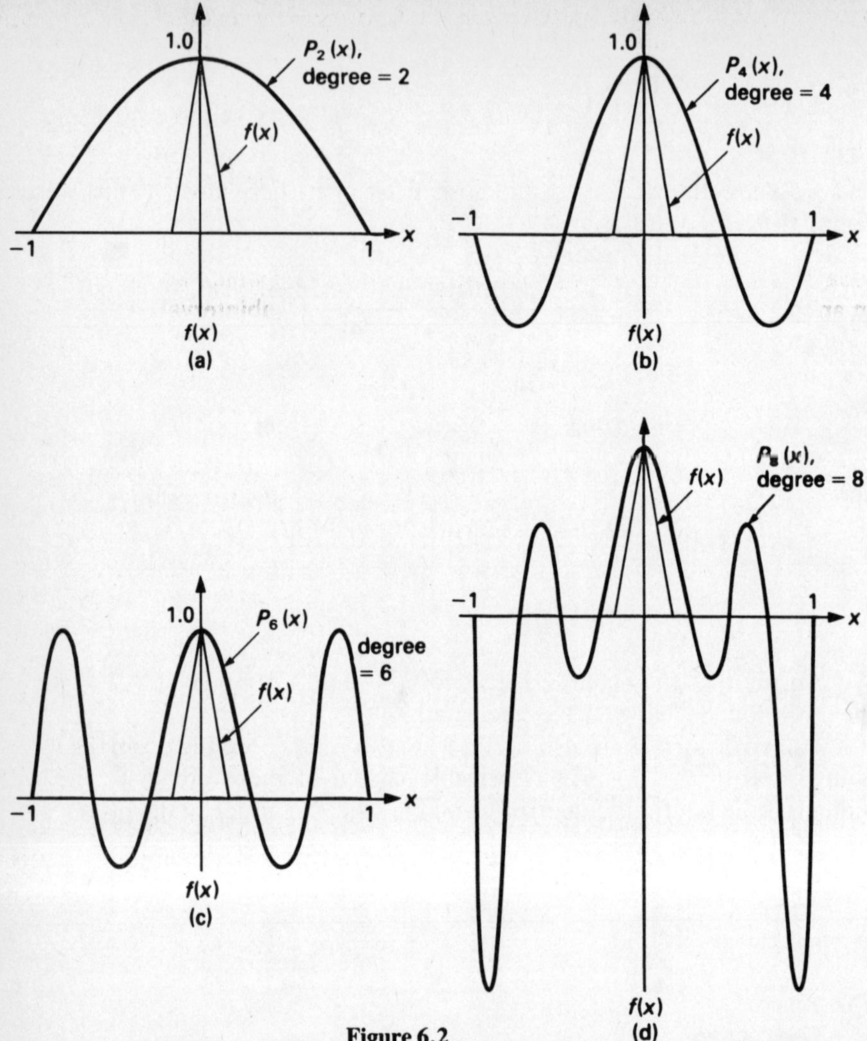

Figure 6.2

One disadvantage of the Lagrange form (there are other types) of interpolation polynomial is that all polynomial values, $l_i(x)$, have to be recalculated if an extra value is provided. There are modifications of the Lagrange polynomial which circumvent this problem and, for the case of equidistant data, extremely efficient methods exist to evaluate the polynomial (see Ralston and Rabinowitz, 1978, p.56). However, polynomial interpolation suffers from a more grave disadvantage which is nicely illustrated by an example in Gerald (1978, p.474). In this example we are trying to interpolate the continuous function

$$f(x) = \begin{cases} 0 & -1 \leq x \leq -0.2 \\ 5(0.2 - |x|) & -0.2 \leq x \leq 0.2 \\ 0 & 0.2 \leq x \leq +1 \end{cases}$$

The diagram of Fig. 6.2 illustrates the approximations obtained by using interpolation polynomials of degrees 2, 4, 6 and 8 on a set of equally spaced data points.

We can see that when the degree is high the interpolation polynomial has large oscillations which are distant from the bump in the original function. The existence of oscillations gives a strong argument for not using interpolation on data which, although smooth, has a local roughness.

The fact that the roughness in this example is localised gives rise to the idea of having local approximations; the interval over which we are interested in approximating the data is split up into a series of subintervals and separate approximations are used in each subinterval. This is subject to the proviso that the approximations should match up, in some sense, at the ends of the subintervals. The simplest form of this type of approximation is illustrated in Fig. 6.3 in which the function is approximated by a straight line in each subinterval; the function values at the ends of the subintervals coincide so that the overall approximation is continuous. This is the type of interpolation that is used with logarithmic tables to evaluate logarithms at non-tabular points.

This process of approximation on subintervals is known, for obvious reasons, as piecewise approximation. Unfortunately, as we can see from the diagram, this approximation to the function is not 'smooth' ('smoothness' usually refers to the continuity of the derivatives) because, at the end-points (sometimes known as *nodes*) of each subinterval the derivative of the approximation is discontinuous. We can try to make the approximation smoother by using piecewise quadratic, rather than piecewise linear, approximation. A quadratic has three free parameters, two of which are determined by the function values at the ends of the subinterval, leaving the third free to be used to smooth the approximation. Unfortunately, there are not enough

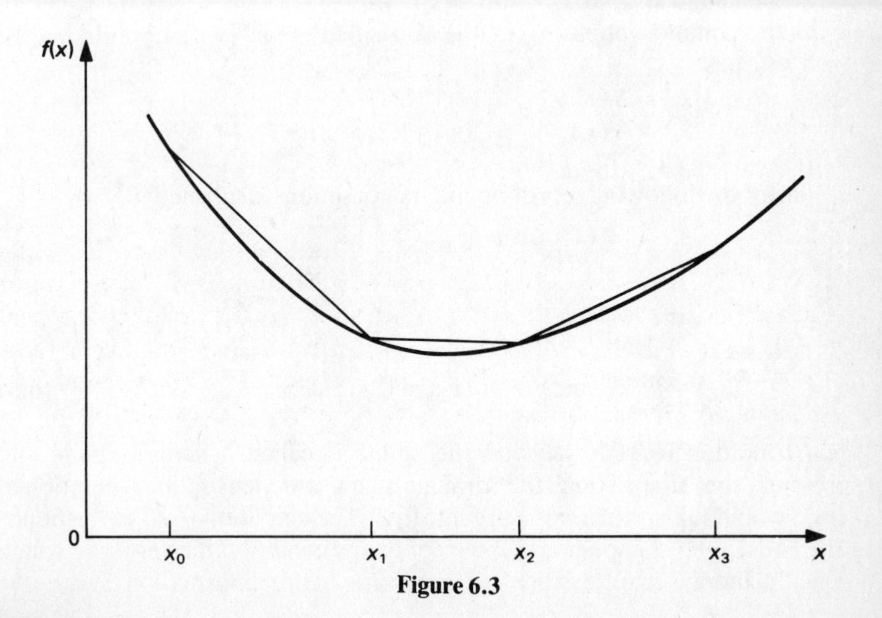

Figure 6.3

free parameters to ensure smoothness over the whole interval; the approximation cannot match the derivatives of the function at the end-points of the interval. This can be achieved, however, by using cubic polynomials in each subinterval and then the approximating polynomials are known as *cubic splines*.

6.1.2 Cubic spline interpolation

A 'spline' is an instrument once used by draughtsmen to join points on a curve. It consists of a flexible wooden strip which was forced to match up with the data points and, because the spline would take up the shape which minimised the potential energy, the resulting curve was smooth. (A jagged curve corresponded to a broken spline!) We will say a little more about the wooden form of the spline later.

Cubic splines have the advantage of sufficient free parameters to ensure continuity of first *and* second derivatives throughout the interval and to satisfy a derivative condition at the ends of the interval. It is important to note that the construction of a cubic spline does not assume that the derivatives of the interpolant agree with those of the function anywhere except, perhaps, at the ends of the interval.

Definition 6.1

Given the set of function values $f_0,f_1,\ldots,f_n$ defined at the distinct points $x_0,x_1,\ldots,x_n$, a cubic spline interpolant S for the data satisfies the following conditions:

(1) in each interval $[x_i,x_{i+1}]$ $\{i=0,1,\ldots,n-1\}$ S is a cubic polynomial denoted by S_i
(2) at each point x_i $\{i=0,1,\ldots,n\}$ $S(x_i)=f_i$
(3) at the common point, x_i, of the two subintervals $[x_{i-1},x_i]$ and $[x_i,x_{i+1}]$ S satisfies

$$\left.\begin{array}{ll}\text{(i)} & S_{i-1}(x_i) = S_i(x_i)\\ \text{(ii)} & S'_{i-1}(x_i) = S'_i(x_i)\\ \text{(iii)} & S''_{i-1}(x_i) = S''_i(x_i)\end{array}\right\}\quad \text{for } i=1,2,\ldots,n-1$$

(4) one of the following sets of boundary conditions is satisfied:

$$S''(x_0) = S''(x_n) = 0 \tag{6.5}$$

or

$$S'(x_0) = f'_0 \quad \text{and} \quad S'(x_n) = f'_n \tag{6.6}$$

When condition (6.5) is satisfied the spline is called a *natural spline* and represents the shape that the draughtman's wooden spline, mentioned earlier, would take if the ends were left free. The condition (6.6) corresponds to the ends being clamped, and if we apply this condition the resulting spline is usually more accurate since we are using more information about the function. However, this information may not be available, in which case either we must take approximations to the derivatives or we must apply the condition (6.5) instead.

Theorem 6.5

Let $f(x)$ be a function defined on the interval $[x_0,x_n]$. Then

(i) $f(x)$ has a unique natural spline interpolant satisfying condition (6.5), and
(ii) $f(x)$ has a unique cubic spline interpolant satisfying condition (6.6).

A proof can be found in Burden *et al.* (1981, pp.111–113).

We now consider how to evaluate the spline for the given set of data. In each subinterval $[x_i,x_{i+1}]$ the spline has the form

$$S_i(x) = a_i(x-x_i)^3 + b_i(x-x_i)^2 + c_i(x-x_i) + d_i \qquad \text{for } i=0,1,\ldots,n-1 \tag{6.7}$$

Now it is obvious that, at x_i,

$$S_i(x_i) = d_i = f_i \qquad \text{for } i=0,1,\ldots,n-1 \tag{6.8}$$

Similarly, when we apply condition 3(i), from Definition 6.1, we obtain

$$\begin{aligned} d_i &= S_i(x_i) = S_{i-1}(x_i) \\ &= a_{i-1}(x_i-x_{i-1})^3 + b_{i-1}(x_i-x_{i-1})^2 + c_{i-1}(x_i-x_{i-1}) + d_{i-1} \\ &\qquad \text{for } i=1,2,\ldots,n-1 \end{aligned} \tag{6.9}$$

The expression (x_i-x_{i-1}) will be repeated frequently and so, to simplify the notation, we define

$$h_{i-1} = x_i-x_{i-1} \qquad \text{for } i=1,2,\ldots,n$$

Thus, from (6.9), upon defining $d_n=S_{n-1}(x_n)=f_n$, we can see that

$$d_i = a_{i-1}h_{i-1}^3 + b_{i-1}h_{i-1}^2 + c_{i-1}h_{i-1} + d_{i-1} \qquad \text{for } i=1,2,\ldots,n \tag{6.10}$$

By differentiating (6.7) once and setting $x=x_i$, we see that

$$S_i'(x_i) = c_i \qquad \text{for } i=0,1,\ldots,n-1$$

and so, by considering condition 3(ii) of Definition 6.1 and defining $c_n=S_{n-1}'(x_n)$, we have that

$$c_i = 3a_{i-1}h_{i-1}^2 + 2b_{i-1}h_{i-1} + c_{i-1} \qquad \text{for } i=1,2,\ldots,n \tag{6.11}$$

In a similar manner the relationships

$$b_i = S_i''(x_i)/2 \qquad \text{for } i=0,1,\ldots,n-1$$

and

$$b_i = 3a_{i-1}h_{i-1} + b_{i-1} \qquad \text{for } i=1,2,\ldots,n \tag{6.12}$$

can be found, where b_n is defined as $b_n = S_{n-1}''(x_n)/2$.

We work backwards from (6.12); solving this equation for a_{i-1} gives

$$a_{i-1} = (b_i - b_{i-1})/3h_{i-1} \qquad \text{for } i=1,2,\ldots,n$$

which can be substituted into equations (6.10) and (6.11) to give

$$d_i = (b_i + 2b_{i-1})h_{i-1}^2/3 + c_{i-1}h_{i-1} + d_{i-1} \qquad \text{for } i=1,2,\ldots,n \tag{6.13}$$

and

$$c_i = (b_i + b_{i-1})h_{i-1} + c_{i-1} \qquad \text{for } i=1,2,\ldots,n \tag{6.14}$$

respectively. Finally we rearrange (6.13) to give

$$c_{i-1} = (d_i - d_{i-1})/h_{i-1} - (b_i + 2b_{i-1})h_{i-1}/3 \qquad \text{for } i=1,2,\ldots,n \tag{6.15}$$

and, increasing the index by one

$$c_i = (d_{i+1} - d_i)/h_i - (b_{i+1} + 2b_i)h_i/3 \qquad \text{for } i=0,1,\ldots,n-1$$

which we now substitute into equation (6.14) to give

$$b_{i-1}h_{i-1} + 2b_i(h_i + h_{i-1}) + b_{i+1}h_i \tag{6.16}$$

$$= \frac{3}{h_i}(d_{i+1} - d_i) - \frac{3}{h_{i-1}}(d_i - d_{i-1}) \qquad \text{for } i=1,2,\ldots,n-1$$

The only unknowns in equation (6.16) are the values b_{i-1}, b_i and b_{i+1} because the h_i-values are predetermined by the x_i-values and, from (6.8), the d_i-values are the nodal values of f. It can be shown that the linear equations (6.16), using either of the boundary conditions (6.5) or (6.6), can be solved for the unknown values b_i $\{i=0,1,\ldots,n\}$. It would appear that there are $(n-1)$ equations in $(n+1)$ unknowns but remember that the conditions (6.5) or (6.6) provide two extra equations. When the values for the b_i $\{i=0,1,\ldots,n\}$ have been obtained, then the a_i-values can be obtained using (6.12). The value c_0 is then obtained from (6.15) and then the remaining c_i-values from (6.14).

The procedure in Fig. 6.4 determines the coefficients of the cubic spline passing through the data values (x_i, f_i) $\{i=0,1,...,n\}$ and satisfying the boundary condition (6.5). This boundary condition implies that $b_0=b_n=0$ and so (6.16) represents a system of $(n-1)$ tri-diagonal equations in the $(n-1)$ unknowns b_i $\{i=1,2,...,n-1\}$. The tri-diagonal matrix is strictly diagonally dominant and as a result is well-conditioned, and no zero pivot can arise during the elimination. The procedure assumes the following environment

```
const
  n = . . .;

type
  subs = 0 . . n;
  rows = record
           a, b, c, d : real
         end {rows};
  rowvectors = array [subs] of rows;
  vectors = array [subs] of real;
```

The procedure *Tridiag* is essentially as in Fig. 4.10 except that the size of the system (n in Fig. 4.10) is passed as a parameter and no check is made for zero pivots (and hence the parameter *success* of Fig. 4.10 is not needed).

```
procedure CubicSplines (h, f : vectors;   var s : rowvectors);
  var
    i : subs;
    eqn : rowvectors;
    bb, cc, hh, ff, fi, hi, nextf : real;
    bvec : vectors;
begin
  hh := h[0];  ff := f[0];  nextf := f[1];
  for i := 1 to n - 1 do
  with eqn[i] do
  begin
    fi := nextf;  nextf := f[i + 1];  hi := h[i];
    b := 3 * ((nextf - fi)/hi - (fi - ff)/hh);
    d := 2 * (hh + hi);  a := hh;  c := hi;
    hh := hi;  ff := fi
  end;
```

```
    Tridiag (eqn, n - 1, bvec);

    bb := 0;   hh := h[0];
    with s[0] do
    begin
      b := bb;   d := f[0];
      cc := (f[1] - d)/hh - bvec[1] * hh/3;
      c := cc
    end;
    for i := 1 to n - 1 do
    with s[i] do
    begin
      b := bvec[i];   d := f[i];
      cc := (b + bb) * hh + cc;   c := cc;
      s[i - 1].a := (b - bb)/(3 * hh);
      bb := b;   hh := h[i]
    end;
    s[n - 1].a := -bb/(3 * hh)
  end { Cubic splines };
```

Figure 6.4

The function of Fig. 6.2 was evaluated at 13 equally spaced points and the function *Lagrange* and procedure *CubicSplines* were applied to the task of estimating the function value at $\bar{x}=0.75$. The estimate produced using *Lagrange* was $p_{12}(\bar{x})=-0.97925$ and, as expected, is not very accurate. In contrast the estimate produced using *CubicSplines* was $S(\bar{x})=-0.00505$ which is much better.

When condition (6.6) is imposed, then $c_0=c_n=0$; this means that (6.16) *is* a system of $(n-1)$ equations in the $(n+1)$ unknowns b_i $\{i=0,1,...,n\}$. However, two extra equations can be obtained from the boundary conditions. At $x=x_0$, $c_0=0$ and (6.15), with $i=1$, gives one condition involving b_0 and b_1. At $x=x_n$ (6.14) is evaluated for $i=n$ and the value c_{n-1} is replaced using (6.15), with $i=n$, giving an equation linking b_{n-1} and b_n. It is left to the reader to modify the procedure for the case when condition (6.6) is used instead.

A modification of the technique described above, which removes the necessity to specify extra boundary conditions, is to choose $x_1, x_2, ..., x_{n-1}$ only as the spline nodes and to specify that

$$S(x_0) = f_0 \qquad \text{and} \qquad S(x_n) = f_n$$

In this case, of course, we have no idea about the continuity of the spline at the nodes x_1 and x_{n-1} but, nevertheless, this method works well in many cases where information about the derivatives of f is not available.

The analysis of the accuracy of cubic spline interpolation is beyond the scope of this book but the following fundamental result can be found in the paper by Birkhoff and de Boor (1964).

Theorem 6.6

If the function $f(x)$ has a continuous *fourth* derivative and we define h=max (h_i) and max (h/h_i) is bounded as $h \to 0$, then

$$\max_{x_0 \leq x \leq x_n} | f^{(k)}(x) - S^{(k)}(x) | \leq Ch^{4-k} \qquad \text{for } k=0,1,2$$

as $h \to 0$, for some constant C independent of h.

Consequently the difference between the function and the cubic spline, at a fixed point x, behaves as h^4 as h tends to zero.

We shall meet splines again in Chapter 7 when we discuss methods for evaluating derivatives of functions. Now, however, we continue by discussing methods for fitting data which is subject to (possibly experimental) error.

6.2 Data fitting

The situation that we shall discuss in this section can be easily illustrated using the diagram in Fig. 6.5. The data values, denoted by crosses in the diagram, have been obtained from some experiment. The underlying theory of the experiment suggests that there should be a linear relationship

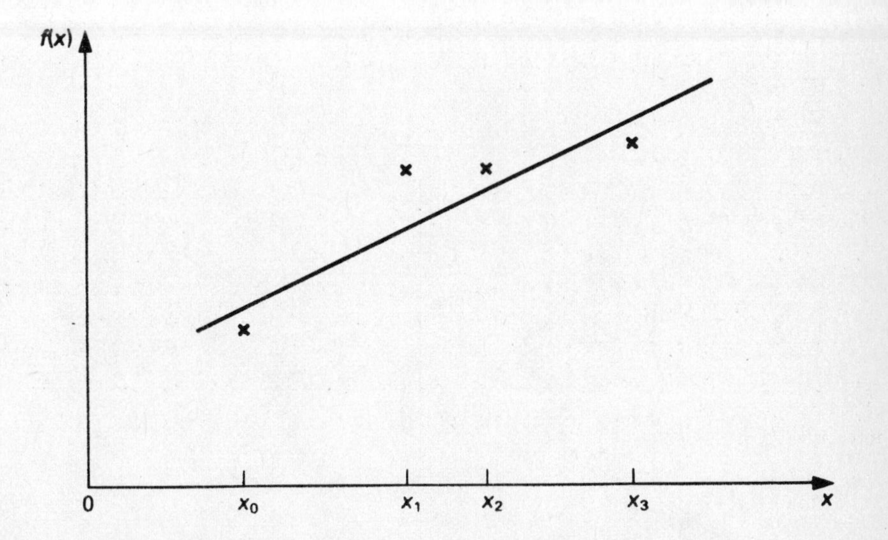

Figure 6.5

$$y = a_1x + a_0$$

between the x and y values. As a result of the errors, which occurred when readings were taken during the experiment, the points do *not* lie on a straight line. However, the experimenter has faith in his theory and wants to know which straight line 'best' fits the data that he has provided. There are now two problems to solve. Firstly we must decide what 'best' means and having made that decision we must then find a method of calculating this 'best' fit.

In this section, we shall concentrate on finding best polynomial approximations to the data. Later, there will be a brief discussion on the use of different types of approximating function.

We now address ourselves to answering the first question: 'What do we mean by best?' Really, what we are looking for is some measure of the difference between the data and the values of the approximating polynomial. For example, if the data (x_0,f_0), (x_1,f_1),. . ., (x_m,f_m) is given and we want to use a polynomial of degree n $(n<m-1)$

$$p_n(x) = a_nx^n + a_{n-1}x^{n-1} + \cdots + a_1x + a_0$$

to fit this data, then we are interested in the *residuals*

$$f_i - p_n(x_i) \qquad \text{for } i=0,1,\ldots,m \tag{6.17}$$

We have already come across measures of differences in Chapter 4 where, specifically, we were looking at differences between vectors. The norms discussed there are relevant to the problem we now have; for example, the L_1-norm of our set of function values is given by

$$\sum_{i=0}^{m} |f_i|$$

the L_2-norm by

$$\left(\sum_{i=0}^{m} |f_i|^2\right)^{1/2}$$

and the L_∞-norm by

$$\max_i |f_i|$$

Indeed we can think of the values f_i $\{i=0,1,\ldots,m\}$ as elements of one vector and, similarly, the values of our approximating polynomial

$$p_n(x_i) = \sum_{j=0}^{n} a_j x_i^j \qquad \text{for } i=0,1,\ldots,m$$

as elements of another. Thus the quantity of interest is a familiar one; the difference of two vectors.

The measure most commonly used to solve our problem is the L_2-norm. One of the major reasons for this is that the calculation of the 'best' L_2-fit is simpler than that for either the L_1-fit or the L_∞-fit. There is also a theoretical reason for preferring the L_2 or *least squares* approximation; the L_1-approximation, which effectively averages the difference between the two vectors, places too little weight on a value that is considerably different from the approximation, the L_∞-fit places too much weight on a similar point and it is felt that, of the three, the L_2-fit places sufficient emphasis on such a point without allowing it to dominate the approximation.

6.2.1 Least squares approximation

In the least squares method we choose the unknown coefficients, $a_0,a_1,\ldots,a_n$, of the polynomial to minimise the L_2-norm of the differences (6.17); that is, we minimise the function

$$F_2 = \left[\sum_{i=0}^{m} \left(f_i - \sum_{j=0}^{n} a_j x_i^j \right)^2 \right]^{1/2} \tag{6.18}$$

with respect to the variables a_j $(j=0,1,\ldots,n)$. In fact, we minimise F_2^2 because this function has a minimum at the same point as F_2 but we avoid the need to calculate the square root. Now, F_2^2 is a continuous function of the unknowns and so we can identify its minimum by setting its first derivatives to zero. (We can make F_2^2 arbitrarily large by an appropriate choice of the a_j $\{j=0,1,\ldots,n\}$ and so, because of the form of F_2^2, zero first derivatives are associated only with minima.) When we differentiate F_2^2 with respect to an arbitrary coefficient a_k and set the result to zero we obtain the equation

$$2\sum_{i=0}^{m} \left(f_i - \sum_{j=0}^{n} a_j x_i^j \right) (-x_i^k) = 0$$

which we can rearrange to give

$$\sum_{i=0}^{m} \sum_{j=0}^{n} a_j x_i^{j+k} = \sum_{i=0}^{m} f_i x_i^k \tag{6.19}$$

Now, since the choice of the coefficient a_k was arbitrary, equation (6.19) must hold for each $k=0,1,...,n$, resulting in a system of $n+1$ simultaneous linear equations in the $n+1$ unknowns a_k $\{k=0,1,...,n\}$. These equations can be rewritten in the obvious matrix–vector form

$$\boldsymbol{Xa} = \boldsymbol{b} \tag{6.20}$$

where

$$x_{jk} = \sum_{i=0}^{m} x_i^{j+k} \qquad \text{for } j,k=0,1,...,n$$

and

$$b_k = \sum_{i=0}^{m} f_i x_i^k$$

and have a unique solution provided that the x_i are distinct. It now appears that a solution to (6.20) can be obtained very easily by applying any of the methods of Section 4.1 and, in fact, this *is* the case when n, the degree of the polynomial, does not exceed 5 or 6, say. For larger values of n the least squares approximations obtained by solving (6.19) become progressively worse, independently of the method of solution. This happens because the matrix X is closely related to the ill-conditioned Hilbert matrix, mentioned in Exercise 9 of Chapter 4. It will not be possible for us to restrict ourselves to using polynomial approximations of low degree only; some way round this problem must be found.

One solution is to define a sequence of polynomials $P_j(x)$ of degree j $\{j=0,1,...\}$ which are orthogonal on the set of data points $\{x_i\}$. In this case two polynomials $P_k(x)$ and $P_l(x)$ are said to be orthogonal if

$$\sum_{i=0}^{m} P_k(x_i)P_l(x_i) \begin{cases} = 0 & k \neq l \\ > 0 & k = l \end{cases}$$

Now, instead of using the polynomial

$$\sum_{j=0}^{n} a_j x^j$$

as the approximation, we use

$$\sum_{j=0}^{n} b_j P_j(x)$$

Theorem 6.7

If the polynomials $P_j(x)$ $\{j=0,1,\ldots,n\}$ of degree j are orthogonal on the set of points $\{x_i\}$ $\{i=0,1,\ldots,m\}$, then the least squares approximation to $f(x)$, on the set of points $\{x_i\}$, by a polynomial of degree at most n is given by

$$p_n(x) = \sum_{j=0}^{n} b_j P_j(x)$$

where

$$b_j = \frac{\sum_{i=0}^{m} f_i P_j(x_i)}{\sum_{i=0}^{m} P_j^2(x_i)}$$

Proof

The function F_2^2 to be minimised now has the form

$$F_2^2 = \sum_{i=0}^{m} \left[f_i - \sum_{j=0}^{n} b_j P_j(x_i) \right]^2$$

Differentiating with respect to an arbitrary coefficient b_k leads to the normal equations

$$\sum_{i=0}^{m} \sum_{j=0}^{n} b_j P_j(x_i) P_k(x_i) = \sum_{i=0}^{m} f_i P_k(x_i) \qquad \text{for } k=0,1,\ldots,n$$

or, changing the order of the summation,

$$\sum_{j=0}^{n} b_j \sum_{i=0}^{m} P_j(x_i) P_k(x_i) = \sum_{i=0}^{m} f_i P_k(x_i) \qquad \text{for } k=0,1,\ldots,n$$

Now, using the orthogonality of the $P_j(x)$, we have

$$b_k \sum_{i=0}^{m} P_k^2(x_i) = \sum_{i=0}^{m} f_i P_k(x_i) \qquad \text{for } k=0,1,\ldots,n$$

from which the result follows.

The disadvantage of this approach is that whenever the data points, x_i, are changed, then so are the equations of the orthogonal polynomials; in this sense they are similar to the Lagrange polynomials.

Another approach involves reformulating the problem. If we return to the original form for the approximating polynomial

$$p(x) = \sum_{j=0}^{n} a_j x^j$$

and now insist that, at each of the data points x_i, the approximating polynomial *interpolates* the data, then we arrive at the set of equations

$$p(x_i) = \sum_{j=0}^{n} a_j x_i^j = f_i \qquad \text{for } i=0,1,\ldots,m \tag{6.21}$$

But the original specification of the problem was that m was larger than $n+1$ so that (6.21) is a *rectangular* system of m equations in n unknowns. In general, such a system of equations does not have a solution; a solution can only be found if the data *can* be interpolated by a polynomial of degree n. However, if we rewrite (6.21) in the form

$$\boldsymbol{Ba} = \boldsymbol{f}$$

where $b_{ij}=x_i^j$, then an approximate solution can be obtained by minimising the square of the L_2-norm of the residuals

$$\| \boldsymbol{Ba} - \boldsymbol{f} \|_2^2 \tag{6.22}$$

If we write this function out in full we obtain

$$\| \boldsymbol{Ba} - \boldsymbol{f} \|_2^2 = \sum_{i=0}^{m} \left(f_i - \sum_{j=0}^{n} a_j x_i^j \right)^2$$

which is exactly the function F_2^2 of equation (6.18) and so the two approaches are equivalent. They differ, however, in their method of solution. We are now going to use the similarity transformations, described in Chapter 5, to minimise (6.22). First of all we note that, if H is an orthogonal matrix, then

$$\| H(\boldsymbol{Ba} = \boldsymbol{f}) \|_2^2 = \| \boldsymbol{Ba} - \boldsymbol{f} \|_2^2$$

Now, from a generalisation of the result used as the basis for the QR algorithm, we know that an orthogonal matrix H can be found, with the property that it 'triangularises' the $m \times n$ matrix B. If we partition B into an $n \times n$ block R and an $(m-n) \times n$ block S so that

$$B = \begin{bmatrix} R \\ S \end{bmatrix}$$

then the result of multiplying B by H is to give

$$HB = \begin{bmatrix} R^* \\ 0 \end{bmatrix}$$

where R^* is an upper triangular $n \times n$ matrix. Also

$$H\boldsymbol{f} = \boldsymbol{c} = \begin{bmatrix} \boldsymbol{c}_1 \\ \boldsymbol{c}_2 \end{bmatrix}$$

where $\boldsymbol{c}_1$ is an n-vector and $\boldsymbol{c}_2$ is an $(m-n)$-vector. As a result of this transformation we are now minimising the function

$$\| H(B\boldsymbol{a} - \boldsymbol{f}) \|_2^2 = \left\| \begin{bmatrix} R^*\boldsymbol{a} - \boldsymbol{c}_1 \\ -\boldsymbol{c}_2 \end{bmatrix} \right\|_2^2 = \| R^*\boldsymbol{a} - \boldsymbol{c}_1 \|_2^2 + \| \boldsymbol{c}_2 \|_2^2 \qquad (6.23)$$

instead of (6.22). The vector $\boldsymbol{a}$ which minimises (6.23) is obviously that which solves the equation

$$R^*\boldsymbol{a} = \boldsymbol{c}_1$$

exactly. The solution of this last equation is particularly simple since R^* is an upper triangular matrix and so only a back-substitution is required. Since $R^*\boldsymbol{a}=\boldsymbol{c}_1$, the minimum value of (6.22) is $\| \boldsymbol{c}_2 \|_2^2$.

The matrix H is composed of the product of $(n-1)$ Householder transformation matrices, similar to those described in Chapter 5, each of which reduces one column of B, below the diagonal, to zero and produces the final matrix R^*. It is possible to show that the system of equations

$$R^*\boldsymbol{a} = \boldsymbol{c}_1$$

is better conditioned than the system (6.20) and so this method of solution is to be preferred.

The procedure of Fig. 6.6 outlines the whole reduction process and, in order to improve efficiency, operates on the transpose of a matrix A so that slices of A may be used. It assumes the following environment

```
const
  n = . . .;   m = . . .;   {m>n+1}
type
  nsubs = 0 . . n;
  msubs = 0 . . m;
  nvectors = array [nsubs] of real;
  mvectors = array [msubs] of real;
```

```
procedure Reduce (var at : nbym;    var f : mvectors);

  var
    i, k : msubs;
    j : nsubs;
    u : real;
    w : mvectors;

  function dotprod (u, v : mvectors;    k : nsubs) : real;
    var
      i : msubs;
      dot : real;
  begin
    dot := 0;
    for i := k to m do
      dot := dot + u[i] * v[i];
    dotprod := dot
  end { dot prod };

  procedure Formw (var w, atk : mvectors;    k : nsubs);
      {Forms w_k . . . w_m and updates a^T_kk}
    var
      i : msubs;   atkk, s, alpha, sumsq : real;
  begin
    sumsq := 0;   atkk := atk[k];
    for i := k to m do
      sumsq := sumsq + sqr(atk[i]);
    if atkk < 0 then s := - sqrt(sumsq)
                else s := + sqrt(sumsq);
    alpha := sqrt (2 * s * (s + atkk));
```

```
      w[k] := (atkk + s) / alpha;   atk[k] := -s;
      for i := k + 1 to m do
        w[i] := atk[i]/alpha
    end { Form w };

    procedure Rowofat
      (var atj : mvectors;   k : nsubs;   w : mvectors;   twowatj : real);
      var
        i : msubs;
    begin
      for i := k to m do
        atj[i] := atj[i] - w[i] * twowatj
    end { Row of at };

  begin { Reduce }
    for k := 0 to n do
    begin
      Formw (w, at[k], k);
      u := dotprod (w, f, k);
      for j := k + 1 to n do
        Rowofat (at[j], k, w, 2 * dotprod (w,at[j],k));
      for i := k to m do
        f[i] := f[i] - 2 * u * w[i]
    end
  end { Reduce };
```

Figure 6.6

Table 6.2

x	-2	-1	0	1	2
$f(x)$	0.1353	0.3679	1.0000	2.7183	7.3891

The procedure of Fig. 6.7 uses this procedure and implements the full least squares process. Given a set of $m + 1$ data values (x_0,f_0), (x_1,f_1),. . ., (x_m,f_m) the procedure determines the coefficients $a_0,a_1,. . .,a_n$ of the least squares polynomial of degree n and computes F_2^2. Table 6.2 contains values of e^x in the range $[-2,2]$ and quoted to four decimal places. When applied to this data and with $n=1$ the procedure of Fig. 6.7 produces the straight line

$$1.6858x + 2.3221$$

```
procedure LeastSquares
    (x, f : mvectors;
    var a : nvectors;   var f2squared : real);

  type
    nbym = array [nsubs] of mvectors;
  var
    i : msubs;   j : nsubs;
    s, xi, xitothej : real;
    bt : nbym;
begin {Least squares}
  {Set up B^T}
  for i := 0 to m do
  begin
    xi := x[i];   xitothej := 1;   bt[0,i] := 1;
    for j := 1 to n do
    begin
      xitothej := xitothej * xi;   bt[j, i] := xitothej
    end
  end;

  Reduce (bt, f);

  {Perform back substitution}
  a[n] := f[n]/bt[n,n];
  for i := n-1 downto 0 do
  begin
    s := 0;
    for j := i + 1 to n do
      s := s + bt[j,i] * a[j];
    a[i] := (f[i] - s) / bt[i,i]
  end;

  {Sum f[n + 1]^2 + · · · + f[m]^2}
  s := 0;
  for i := n + 1 to m do
    s := s + sqr(f[i]);
  f2squared := s
end { Least squares };
```

Figure 6.7

We should note that there is an alternative form of least squares approximation: the *weighted* least square approximation. In this method we minimise the functional

$$W_2 = \left[\sum_{i=0}^{m} \omega_i \left(f_i - \sum_{j=0}^{n} a_j x_i^j \right)^2 \right]^{1/2} \tag{6.24}$$

instead of (6.18). In this function the *weights* ω_i $\{i=0,1,\ldots,m\}$ are positive values that remain to be chosen. They are used, in general, to emphasize certain of the residuals. In this way we would attach a large weight to residuals that we particularly wanted to be small in the final solution. The least squares technique will 'see' these residuals as being large and will calculate a solution to compensate for this.

The matrix–vector form of W_2^2, in (6.24), is obtained by multiplying the vector $(Ba-f)$ by the $m \times m$ diagonal matrix W with the weights, in order, on the diagonal. The solution technique is then as for (6.22).

6.2.2 Minimax approximation

The least squares approximation suffers from the disadvantage that an estimate of only the total error in the approximation is available; there is no estimate for the accuracy of the approximation at each of the data points. The *minimax* (or *Chebyshev*, or *uniform*) approximation does provide such an error bound. It makes use of the L_∞-norm of the residuals; the coefficients a_j $\{j=0,1,\ldots,n\}$ are chosen to minimise the function

$$M_\infty = \max_i \left(\left| f_i - \sum_{j=0}^{n} a_j x_i^j \right| \right) \tag{6.25}$$

whence M_∞ *is* the upper bound on the errors. We can see immediately that minimisation of M_∞ (hence the name *minimax*) will be less straightforward than was that of F_2^2; the function in (6.25) is not differentiable, because of the modulus sign, and so its minimum value must be found using another method.

Theorem 6.8

The minimax polynomial approximation

$$p_n(x) = \sum_{j=0}^{n} a_j x^j$$

to the data f_i $\{i=0,1,\ldots,m\}$ is unique.

The following theorem categorises the minimax polynomial.

Theorem 6.9
The polynomial

$$p_n(x) = \sum_{j=0}^{n} a_j x^j$$

is the minimax polynomial of degree n for the data f_i $\{i=0,1,\ldots,m\}$ if and only if the maximum modulus error

$$\max_i |\epsilon_i| = \max_i(|f_i - p_n(x_i)|) \qquad \text{for } i=0,1,\ldots,m$$

is attained on a subset of *at least* $(n+2)$ points $x_{(0)},x_{(1)},\ldots,x_{(n+1)}$, say, such that

$$\epsilon_{(i)} = -\epsilon_{(i+1)} \qquad \text{for } (i)=(0),(1),\ldots,(n)$$

Proofs of these theorems can be found in Rice (1964, p.52 *et seq.*).

Theorem 6.9 tells us that if we can find a polynomial satisfying the alternating sign property then it must be the minimax polynomial. This alternating error property is the basis of several iterative algorithms for determining the minimax polynomial. An initial guess is made at the set of points $\{x_{(0)},x_{(1)},\ldots,x_{(n+1)}\}$ at which the maximum error is to occur. The following system of linear equations

$$f_{(i)} - \sum_{j=0}^{n} a_j x_{(i)}^j = (-1)^i d \qquad \text{for } (i)=(0),(1),\ldots,(n+1) \qquad (6.26)$$

is solved, using a technique from Chapter 4, for the $n+2$ unknowns a_j $\{j=0,1,\ldots,n\}$ and d. Now the errors, ϵ_i, at all the m points x_i $\{i=0,1,\ldots,m\}$ are evaluated. If any of the ϵ_i is greater in modulus than d then the polynomial is *not* the minimax polynomial and a further iteration must take place; if no ϵ_i is greater than d then the minimax polynomial has been found and the iteration stops. The iteration continues by *exchanging* points in the set $\{x_{(0)},x_{(1)},\ldots,x_{(n+1)}\}$ with points at which the error exceeds d in modulus in such a way that the alternating property of the signs of the errors in the set remains.

Now (6.26) is rewritten in terms of the new set of points and then solved as before. This iterative process continues until the *maximum* error, satisfying the criterion of the theorem, is achieved.

The algorithm we shall describe is known as the *first* exchange algorithm of *Remes* and exchanges only the point with *maximum* error if its modulus exceeds d; other methods exchange several points at one time. If the point with maximum error lies outside the range of the current set of points, it is substituted for one of the end-points of the current set; otherwise it replaces one of its neighbouring points from the set.

The following example illustrates the iteration. We wish to determine the minimax line, $y=a_1x+a_0$, for the values of e^x given in Table 6.2. From Theorem 6.9, we need to find *three* points at which the maximum error is achieved and at which the signs of the errors alternate. Let us choose the first three points $\{-2, -1, 0\}$ as the initial triple. We now solve equations (6.26) in the form

$$\begin{aligned} 0.1353 + 2a_1 - a_0 &= d \\ 0.3679 + a_1 - a_0 &= -d \\ 1.0000 \quad\quad - a_0 &= d \end{aligned}$$

These have the solution $a_1=0.4324$, $a_0=0.9001$ and $d=0.0999$ and so the line given by this calculation is $y=0.4323x+0.9001$. We calculate the errors at the remaining two points and the errors at all the points are given in Table 6.3.

Table 6.3

x	-2	-1	0	1	2
Error	0.0999	−0.0999	0.0999	1.3859	5.6243

We can see immediately that the errors at $x=1$ and $x=2$ are larger than the calculated value for d and so we have not yet found the minimax line. Following the description given above we want to exchange the point with the largest error, $x=2$, for one of the original triple in such a way that the signs of the errors of the triple still alternate. It is obvious in this example that we replace $x=0$ by $x=2$, giving the triple $\{-2, -1, 2\}$. Equations (6.26) now have the form

$$\begin{aligned} 0.1353 + 2a_1 - a_0 &= d \\ 0.3679 + a_1 - a_0 &= -d \\ 7.2891 - 2a_1 - a_0 &= d \end{aligned}$$

and have the solution $a_1=1.8135$, $a_0=2.9718$ and $d=0.7904$. This gives the line $y=1.8135x+2.9718$ and the errors at each point are presented in Table 6.4.

Table 6.4

x	-2	-1	0	1	2
Error	0.7904	−0.7904	−1.9718	−2.0669	0.7904

The error at $x=1$ is the largest in modulus and is larger than the calculated value of d. The sign of the error is negative, so the point to be replaced is $x=-1$ and the triple is now $\{-2, 1, 2\}$. Equations (6.26) now take the form

$$\begin{aligned} 0.1352 + 2a_1 - a_0 &= d \\ 2.7183 - a_1 - a_0 &= -d \\ 7.3891 - 2a_1 - a_0 &= d \end{aligned}$$

and have the solution $a_1=1.18135$, $a_0=2.3335$ and $d=1.4287$ giving the line $y=1.8135x+2.3335$. The resulting errors are shown in Table 6.5.

Table 6.5

x	-2	-1	0	1	2
Error	1.4287	−0.1522	−1.3335	−1.4287	1.4287

None of the errors in Table 6.5 exceeds, in modulus, the calculated value of d and so we have found the minimax line; the best minimax fit to the data by a straight line is given by

$$y = 1.8135x + 2.3335$$

The diagram of Fig. 6.8 shows the data of Table 6.2 and both the least squares best straight line (——) and the minimax best straight line (----). Table 6.6 gives the values of these two lines at the five data points and shows the errors at these points.

Table 6.6

x	$f(x)$	$LS(x)$	*Error*	$R(x)$	*Error*
−2	0.1353	−1.0495	1.1848	−1.2934	1.4287
−1	0.3679	0.6363	−0.2684	0.5201	−0.1522
0	1.0000	2.3221	−1.3221	2.3335	−1.3335
1	2.7183	4.0079	−1.2896	4.1470	−1.4287
2	7.3891	5.6937	1.6954	5.9604	1.4287

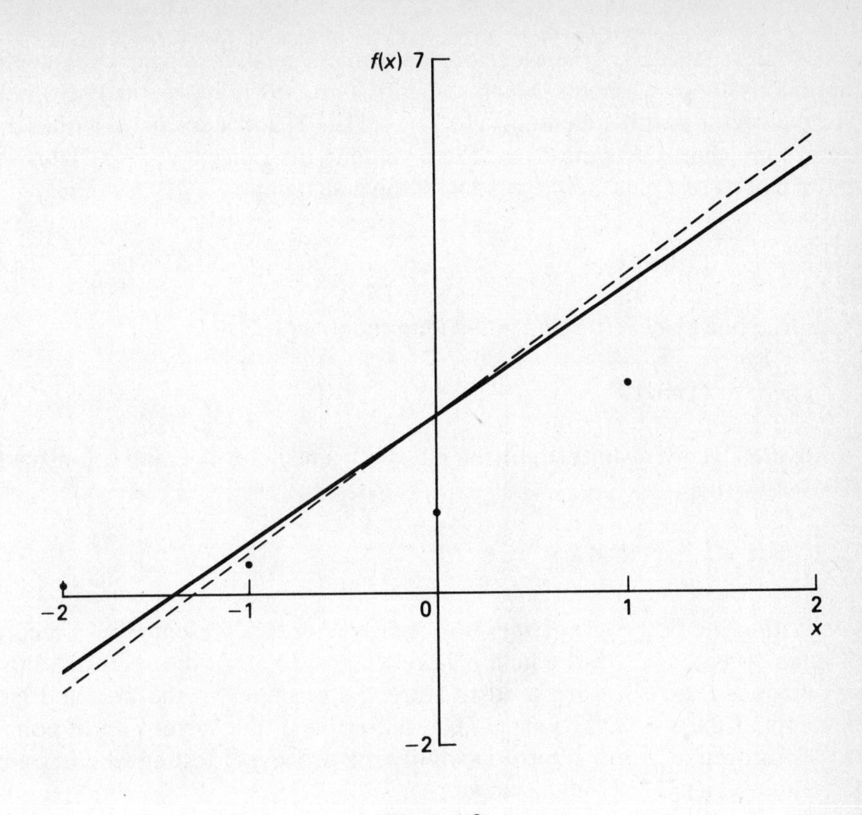

Figure 6.8

The procedure of Fig. 6.9 uses the Remes algorithm described to determine the minimax polynomial of degree n for the $m+1$ data points $(x_0,f_0),\ldots,(x_m,f_m)$, where $m>n+2$. It assumes the following environment

```
const
   n = . . .;   m = . . .;   nplus1 = . . .;   {n + 1}
type
   nsubs = 0 . . n;
   msubs = 0 . . m;
   nplus1subs = 0 . . nplus1;
   nvectors = array [nsubs] of real;
   mvectors = array [msubs] of real;
   nplus1vectors = array [nplus1subs] of real;
   rows = record
              coeff : nplus1vectors;   b : real
          end { rows };
   rowvectors = array [nplus1subs] of rows;
```

The subscripts of the current set of points are stored in a vector. We must be able to add a point at one end and remove one from the other and be able to do this repeatedly. It is therefore convenient to think of the vector as a continuous loop; elements which 'fall off' one end reappear at the other. Given a vector v with k elements $v[0], \ldots, v[k-1]$ and a position i within the vector, any change to i must result in a value in the range 0 to $k-1$. Thus we would move i one place to the right with the assignment

$$i := (i + 1) \textbf{ mod } k$$

To move i one place left we might be tempted to write

$$i := (i - 1) \textbf{ mod } k$$

but we must guard against supplying **mod** with a negative operand so, instead, we would write

$$i := (i + k - 1) \textbf{ mod } k$$

Within the procedure *Remes* this cyclic vector is represented by a record variable *cycvec* in which the field *p* stores the elements of the vector and the two fields *first* and *last* are used to store the positions of the first and last elements of the current sequence. The subscripts of the current set of points are also stored as a set *currpts* to simplify the process of testing whether any given point is a member of the current set.

The procedure *RecordGauss*, used to solve the system of linear equations which are part of each iteration, is essentially as in Fig. 4.4 except that the size of the system (n in Fig. 4.4) is transferred as a parameter and subscripting is modified to run from 0. In general, we do not know whether the system of linear equations, to be solved during each iteration, is well-conditioned; thus the boolean *success* is examined immediately upon return from the call of *RecordGauss* and, if the Gaussian elimination was unsuccessful, the procedure *Remes* is immediately exited.

The function *poly* is a slightly modified version of Fig. 3.13.

```
function poly (a : nplus1vectors;   n : nsubs;   x : real) : real;
  var
    i : nsubs;
    p : real;
begin
  p := a[n];
  for i := n - 1 downto 0 do
    p := p * x + a[i];
  poly := p
end { poly };
```

The two procedures *GetadandErrors* and *UpdateSet* have been introduced essentially to improve readability. They are declared within *Remes* and so can refer directly to the parameters of *Remes* and do not need to take parameters themselves. However, in the interest of program transparency, they have been supplied with variable parameters to draw attention to the variables they change.

```
procedure Remes (x, f : mvectors;
                 var a : nvectors;   var success : boolean);
  type
    ptsets = set of msubs;
    cycptvecs = record
                  p : array [nplus1subs] of msubs;
                  first, last : nplus1subs
                end { cyc pt vecs };
    errors = record
                  e : mvectors;
                max : real;
              maxpt : msubs
             end { errors };

  var
    npts : msubs;
    i : nplus1subs;
    j : nsubs;
    currpts : ptsets;
    err : errors;
    ad : nplus1vectors;
    cycvec : cycptvecs;

  procedure GetadandErrors
        (var ad : nplus1vectors;   var err : errors;
         var solved : boolean);
    {Determines a0, . . ., an and d and computes the error at
     all the points x0, . . ., xm. The ai and d are returned in
     the vector ad (d ≡ ad[n+1]) and the error information in err.
     The errors are stored in the field e and the fields max and
     maxpt store, respectively, the error of maximum modulus
     and the subscript of the point at which this error occurs.}
   var
     eqn : rowvectors;
     dcoeff : -1 . . 1;
     i, before : nplus1subs;
     j : nsubs;
     pt, pbefore : msubs;
     d, error, xi, xitothej : real;
```

```
begin
  dcoeff := -1;   before := cycvec.first;
  for i := 0 to n + 1 do
  with eqn[i] do
  begin
    coeff[0] := 1;
    {Store d as a_{n+1}}
    dcoeff := -dcoeff;   coeff[nplus1] := dcoeff;
    pbefore := cycvec.p[before];   b := f[pbefore];
    xi := x[pbefore];   before := (before + 1) mod npts;
    xitothej := 1;
    for j := 1 to n do
    begin
      xitothej := xitothej * xi;   coeff[j] := xitothej
    end
  end;

  {Solve for a_0, a_1, . . ., a_n and d}
  RecordGauss (eqn, nplus1, ad, solved);

  if solved then
  begin
    {Compute errors at x_0, x_1, . . ., x_m}
    with err do
    begin
      max := 0;   d := -ad[nplus1];
      for pt := 0 to m do
        if pt in currpts then
        begin
          d := -d;   e[pt] := d
        end else
        begin
          error := f[pt] - poly (ad,n,x[pt]);
            e[pt] := error;
            if abs(error) > abs(max) then
            begin
              max := error;   maxpt := pt
            end
          end
        end
    end
  end { Get a d and errors };
```

```
procedure UpdateSet
    (err : errors;   var currpts : ptsets;   var cycvec : cycptvecs);
      {Brings into the current set that point at
       which the error of maximum modulus occurs.}
  var
    pfirst, plast : msubs;
    after, before : nplus1subs;

  procedure Exchange
    (var pts : ptsets;   this : msubs;   var that : msubs);

  begin
    pts := pts + [this] − [that];   that := this
  end { Exchange };

begin {Update set}
  with cycvec, err do
  begin
    pfirst := p[first];   plast := p[last];
    if maxpt < pfirst then
      {exchange with an end point of current set}
    case max * e[pfirst] > 0 of
      true  : Exchange (currpts, maxpt, p[first]);
      false : begin
                 last := (last + nplus1) mod npts;
                 first := (first + nplus1) mod npts;
                 p[first] := maxpt;
                 currpts := currpts + [maxpt] − [plast]
              end
    end { case } else
    if maxpt > plast then
      {exchange with an end point of current set}
    case max * e[plast] > 0 of
      true  : Exchange (currpts, maxpt, p[last]);
      false : begin
                 first := (first + 1) mod npts;
                 last := (last + 1) mod npts;
                 p[last] := maxpt;
                 currpts := currpts + [maxpt] − [pfirst]
              end
    end { case } else
```

```
        begin {exchange with a neighbour from current set}
          before := first;   after := (before + 1) mod npts;
          while maxpt > p[after] do
          begin
            before := after;   after := (after + 1) mod npts
          end;
          if max * e[p[before]] > 0 then
            Exchange (currpts, maxpt, p[before])
          else
            Exchange (currpts, maxpt, p[after])
        end
      end
    end { Update set };


  begin { Remes }
    npts := n + 2;
      {Choose points x0, x1, . . ., xn+1 as current set}
    currpts := [0 . . nplus1];
    with cycvec do
    begin
      first := 0;   last := nplus1;
      for i := first to last do
        p[i] := i
    end;


    GetadandErrors (ad, err, success);
    if success then
    begin
        {Remember, d ≡ ad[n + 1]}
      while success and (abs(err.max) > abs(ad[nplus1])) do
      begin
        UpdateSet (err, currpts, cycvec);
        GetadandErrors (ad, err, success)
      end;
      if success then
        for j := 0 to n do
          a[j] := ad[j]
    end
  end { Remes };
```

Figure 6.9

There is a *second* exchange algorithm of Remes which differs from the first in that *all* points at which the error is greater than the calculated value of d are brought into the set. This is done subject to the proviso that there are not more than $(n+2)$ points in the set and that the alternating sign property is maintained. The exchange algorithms are extremely efficient methods for determining the minimax polynomial. A discussion of their convergence properties is beyond the scope of this book and so we refer the reader to Powell (1981, p.97 *et seq.*) for further details.

The theory behind these algorithms remains the same if *weighted* minimax approximations are required. The L_∞-norm then becomes

$$\max_i(\omega_i \mid f_i \mid)$$

where the ω_i are positive values as before. The exchange algorithms take the same form except that the errors are now weighted values.

6.2.3 Alternative approximating functions

In this chapter we have concentrated on using polynomials to approximate the original data. In many cases this choice is inappropriate. For example, it is frequently the case that the data is periodic (related, say, to the availability of sunlight) or may vary exponentially (many biological processes appear to behave this way). For the first case the use of trigonometric functions seems sensible and these are the basis of the well-known method utilising *Fourier series*; in the second case the use of exponential functions is indicated. A further type of approximating function in common use is the *rational polynomial* consisting of the ratio of two polynomials. Rational polynomials are particularly useful for approximating functions with singularities and we refer the interested reader to Ralston and Rabinowitz (1978).

6.3 Exercises

1. Given that the interpolation points are equally spaced, use the Lagrange error formula to determine at how many points the function $f(x)=e^x$ must be interpolated in order for the resulting polynomial to be accurate to six decimal places everywhere in the interval [0,1].

2. By obtaining numerical values, using the function *Lagrange* of Fig. 6.1, draw the graph of the polynomial which interpolates the data of Table 6.1. Pay particular attention to the intervals [0,0.1] and [0.6,1.0]. Given that the data is derived from the function e^x how accurate do you think that your interpolate is outside the interval [0.1,0.6]?

3. Modify the procedure *CubicSplines* of Fig. 6.4 to use the boundary conditions (6.6). Apply the modified procedure to determine the value, at 0.75, of the interpolate to the data of Fig. 6.2.

4 Derive the normal equations for the least squares line fitting the data of Table 6.1 and using, for example, *PointerGauss* of Fig. 4.5 obtain the coefficients of the polynomial. How do these coefficients compare with those obtained using the procedure *LeastSquares* of Fig. 6.7?

5 Draw the graph of the function e^x on the interval [0,1]. Using the points $x_i=i/10$ $\{i=0,1,\ldots,10\}$ as data points draw, on the same axes, the graphs of the Lagrange polynomial and, using boundary condition (6.5), the cubic spline interpolate. Draw, also, the graphs of the least squares and minimax quadratics for the same data.

6 Modify *LeastSquares* of Fig. 6.7 and *Remes* of Fig. 6.9 to obtain weighted best fits. Apply these procedures to the data of Exercise 5 and note the effect of using differing sets of weights.

Chapter 7 DIFFERENTIATION AND INTEGRATION

There are essentially two situations where numerical differentiation, or integration, is necessary. The first occurs when the data takes the form of discrete values, the same situation as held for the problem discussed in Chapter 6. Here the function values and the points at which they are given are fixed and extra data values may not be available. The second arises when the function to be differentiated, or integrated, is complicated and an analytic differentiation, or integration, is difficult if not impossible; numerical methods can be used instead, always provided that the function itself can be evaluated simply. We consider evaluating definite integrals only. It will be assumed throughout this chapter that the function is sufficiently smooth, without discontinuities or singularities, for the processes which we describe to be meaningful.

7.1 Differentiation

In Chapter 6 we described several methods for obtaining polynomial approximations to functions represented by sets of discrete data. In this section approximations to the derivatives of the function are obtained by differentiating one of the polynomial approximations.

7.1.1 Discrete data

Here we have been given function values at prescribed points and have no access to further information. Because the data points could be arbitrarily spaced it seems reasonable that we should obtain an approximation to the function using Lagrange interpolation and then differentiate the interpolation polynomial. We illustrate this by considering the situation in which we have been given the three sets of values (x_0,f_0), (x_1,f_1) and (x_2,f_2) and we shall assume that the data is always ordered so that $x_0<x_1<\cdots$. The Lagrange interpolation polynomial has degree 2 and is written as

$$\begin{aligned} p_2(x) &= f_0l_0(x) + f_1l_1(x) + f_2l_2(x) \\ &= f_0\,\frac{(x-x_1)(x-x_2)}{(x_0-x_1)(x_0-x_2)} + f_1\frac{(x-x_0)(x-x_2)}{(x_1-x_0)(x_1-x_2)} + f_2\frac{(x-x_0)(x-x_1)}{(x_2-x_0)(x_2-x_1)} \end{aligned}$$

and the error, $f(x)-p_2(x)$, is given by the expression

$$E(x) = \frac{(x-x_0)(x-x_1)(x-x_2)}{3!} f'''(\xi) \qquad x_0<\xi<x_2$$

We now approximate $f'(x)$ by $p_2'(x)$, i.e. by

$$p_2'(x) = f_0 l_0'(x) + f_1 l_1'(x) + f_2 l_2'(x) \tag{7.1}$$

where

$$l_0'(x) = \frac{(2x - x_1 - x_2)}{(x_0-x_1)(x_0-x_2)} \tag{7.2a}$$

$$l_1'(x) = \frac{(2x - x_0 - x_2)}{(x_1-x_0)(x_1-x_2)} \tag{7.2b}$$

$$l_2'(x) = \frac{(2x - x_0 - x_1)}{(x_2-x_0)(x_2-x_1)} \tag{7.2c}$$

The process, therefore, is straightforward; for example, let us try to evaluate the derivative, at $x=1$, of $f(x)=e^x$ given the data values (0.0, 1.0000), (0.5, 1.6487) and (1.5, 4.4817). From equations (7.2) we have

$$\begin{aligned} l_0' &= 0.0000 \\ l_1' &= -1.0000 \\ l_2' &= 1.0000 \end{aligned}$$

and note that $l_0'+l_1'+l_2'=0$; this is a reflection of the fact that the derivative of a constant function is zero. Our approximation to $f'(1)$ is

$$\begin{aligned} p_2'(1) &= 1.0000(0.0000) + 1.6487(-1.0000) + 4.4817(1.0000) \\ &= 2.8330 \end{aligned}$$

The correct value of $f'(1)=e^1$ is 2.7183, to four decimal places, and so the numerical approximation is of the correct order of magnitude but is hardly very accurate. Thus, we must also study the error in the differentiation process in an attempt to see how accurate an answer we can expect.

An extremely convenient feature of differentiation and integration is that they are linear operations; viz.

$$\frac{d}{dx}[f(x) - g(x)] \equiv \frac{d}{dx}[f(x)] - \frac{d}{dx}[g(x)]$$

and

$$\int [f(x) - g(x)]\,dx \equiv \int f(x)\,dx - \int g(x)\,dx$$

Thus, the error in the numerical differentiation is the derivative of the interpolation error. This gives

$$
\begin{aligned}
f'(x) - p_2'(x) &= \frac{d}{dx}[E(x)] \\
&= \frac{d}{dx}\left[\frac{(x-x_0)(x-x_1)(x-x_2)}{3!} f'''(\xi)\right], \qquad x_0<\xi<x_2 \\
&= \frac{1}{3!}\left[\frac{d}{dx}\{(x-x_0)(x-x_1)(x-x_2)\}f'''(\xi)\right. \\
&\qquad \left. + (x-x_0)(x-x_1)(x-x_2)\frac{d}{dx}\{f'''(\xi)\}\right], \qquad x_0<\xi<x_2
\end{aligned}
\tag{7.3}
$$

because the point ξ is a function of the point x at which the interpolate is evaluated. Now, unfortunately, the analysis is less straightforward; we do not know how ξ varies as we change x and so we cannot evaluate the second differential expression in (7.3), leaving us with no idea of the magnitude of the error. However, all is not yet lost; if we concentrate on evaluating the derivative of the interpolation polynomial at one of the data points, then the second term in (7.3) is zero and we can now estimate the error. For example, the function from which our data was obtained is $f(x)=e^x$, whose derivative value at $x_1=0.5$ is 1.6487. The numerical estimate of this derivative is

$$
\begin{aligned}
p_2'(x_1) &= (1.0000)(-1.3333) + (1.6487)(1.0000) + (4.4817)(0.3333) \\
&= 1.8093
\end{aligned}
$$

Using (7.3) we obtain the following upper bound for the error at x_1:

$$
\begin{aligned}
|E'(x_1)| &\le \frac{1}{3!}[3x_1^2 - 2(x_0+x_1+x_2)x_1 + (x_0x_1+x_1x_2+x_2x_0)]e^{1.5} \\
&= 0.3735
\end{aligned}
$$

Similarly, a lower bound is

$$|E'(x_1)| \ge 0.0833$$

and the actual error for this case, 1.8093−1.6487=0.1606, does lie between these bounds.

In general, better results are obtainable by using either cubic splines or the least squares polynomial. Cubic splines are particularly appropriate because their construction is such that the first and second derivatives are continuous and, further, an error estimate is available; it was quoted in Theorem 6.6, p.191.

When the function itself is available and can be evaluated anywhere in a given interval $[a,b]$, then the methods described in the following section are preferable.

7.1.2 Continuous data

In this section we assume that we can choose the number *and* positions of the points at which the function is to be evaluated; then we can usually obtain a more accurate estimate of the derivative. However, there are restrictions and these will be discussed later.

Let us suppose that we want the derivative of the function $f(x)$ at the point $\hat{x}$ and also that the points $\hat{x}+h$ and $\hat{x}-h$ lie inside the interval $[a,b]$. Then, using Taylor's series, we can write

$$f(\hat{x}+h) = f(\hat{x}) + hf'(\hat{x}) + \frac{h^2}{2!}f''(\hat{x}) + \frac{h^3}{3!}f'''(\hat{x}) + \cdots \quad (7.4)$$

and

$$f(\hat{x}-h) = f(\hat{x}) - hf'(\hat{x}) + \frac{h^2}{2!}f''(\hat{x}) - \frac{h^3}{3!}f'''(\hat{x}) + \cdots \quad (7.5)$$

Subtracting (7.5) from (7.4) and dividing by $2h$ results in the equation

$$f'(\hat{x}) = \frac{f(\hat{x}+h) - f(\hat{x}-h)}{2h} - \sum_{j=1}^{\infty} \frac{f^{(2j+1)}(\hat{x})}{(2j+1)!} h^{2j} \quad (7.6)$$

Similarly we can obtain the expression

$$f''(\hat{x}) = \frac{f(\hat{x}-h)-2f(\hat{x})+f(\hat{x}+h)}{h^2} - \sum_{j=1}^{\infty} \frac{f^{(2j+2)}(\hat{x})}{(2j+2)!} h^{2j}$$

As a result, if we choose h small enough, then we can approximate $f'(\hat{x})$ by

$$f'(\hat{x}) = \frac{f(\hat{x}+h) - f(\hat{x}-h)}{2h} \quad (7.7)$$

and $f''(\hat{x})$ by

$$f''(\hat{x}) = \frac{f(\hat{x}-h) - 2f(\hat{x}) + f(\hat{x}+h)}{h^2}$$

Formulae for higher derivatives can be obtained in exactly the same way.

It would seem that we could take a sequence of decreasing positive values of h and obtain arbitrarily good approximations to the first and second derivatives of $f(x)$ at $\hat{x}$. Unfortunately, rounding errors eventually dominate the calculation and we can quite easily see why this should be. If we consider (7.7) and let h get very small in magnitude, then the two terms $f(\hat{x}+h)$ and $f(\hat{x}-h)$ in the numerator must be nearly equal and so their difference must be small. Also, the denominator $2h$ must be small since h is small and so (7.6) consists of the quotient of two numbers each of which is small. In Chapter 2 we advised that this situation be avoided if possible since it was prone to rounding errors.

We can analyse this error further. We rewrite (7.6), using the mean value theorem (Flett 1966, p.189, Ex.5.57) to replace the infinite series for the error, to give

$$f'(\hat{x}) = \frac{f(\hat{x}+h) - f(\hat{x}-h)}{2h} - \frac{h^2}{6} f'''(\xi)$$

where $\hat{x}-h<\xi<\hat{x}+h$. Now, suppose that when we evaluate $f(\hat{x}+h)$ we incur a rounding error $\epsilon(\hat{x}+h)$ and similarly when evaluating $f(\hat{x}-h)$ we incur $\epsilon(\hat{x}-h)$. As a result, the computed values $\tilde{f}(\hat{x}+h)$ and $\tilde{f}(\hat{x}-h)$ have the form

$$\tilde{f}(\hat{x}+h) = f(\hat{x}+h) + \epsilon(\hat{x}+h)$$

and

$$\tilde{f}(\hat{x}-h) = f(\hat{x}-h) + \epsilon(\hat{x}-h)$$

Now the error in our approximation to $f'(\hat{x})$ is

$$f'(\hat{x}) - \frac{\tilde{f}(\hat{x}+h) - \tilde{f}(\hat{x}-h)}{2h} = \frac{\epsilon(\hat{x}+h)-\epsilon(\hat{x}-h)}{2h} - \frac{h^2}{6} f'''(\xi)$$

involving both rounding and truncation errors. If the rounding errors $\epsilon(\hat{x}\pm h)$ are bounded in modulus by some value $\epsilon>0$ and if the third derivative of f is bounded by some value F, then

$$\left| f'(\hat{x}) - \frac{\tilde{f}(\hat{x}+h) - \tilde{f}(\hat{x}-h)}{2h} \right| \leq \frac{\epsilon}{h} + \frac{h^2}{6} F \tag{7.8}$$

If we examine this expression we see that if h is small, then the term ϵ/h can be large; thus the effect of rounding error can dominate the approximation. The values in Table 7.1 represent approximations to the first derivative of $f(x)=e^x$ at the point $x=1$ using (7.8) and a sequence of decreasing values for h. We can see that, initially, the approximation inproves with decreasing h until it is close to the exact value, $f'(\hat{x})=2.7183$, and then diverges. In this case it is possible to estimate the value of h for which the smallest error should occur; the right-hand side of (7.8) is differentiated to determine the point at which a minimum occurs. This point is found to be

$$h = (3\epsilon/F)^{1/3}$$

Table 7.1

h	Approx to $f'(1)$
0.10	2.7230
0.09	2.7222
0.08	2.7213
0.07	2.7207
0.06	2.7200
0.05	2.7200
0.04	2.7188
0.03	2.7200
0.02	2.7175
0.01	2.7200

In our case, because we are working correct to four decimal places, $\epsilon=0.00005$ and we know that F is e^1 and so the optimal value of h is $h \simeq 0.038$, which is consistent with the results above. In general, we will not be able to carry out this calculation for an optimal h because we will not have the information about the third derivative of f.

If it is not possible to obtain an answer to the desired accuracy by decreasing h, then the *Richardson extrapolation* technique can be used. If we expand (7.6) as far as terms in h^4 we obtain

$$f'(\hat{x}) = \frac{f(\hat{x}+h) - f(\hat{x}-h)}{2h} - \frac{h^2}{6} f'''(\hat{x}) + \frac{h^4}{120} f^{v}(\hat{x}) + \cdots \tag{7.9}$$

Now we replace h by rh, where r is any value not equal to ± 1. This results in the equation

$$f'(\hat{x}) = \frac{f(\hat{x}+rh) - f(\hat{x}-rh)}{2rh} - \frac{r^2h^2}{6} f'''(\hat{x}) + \frac{r^4h^4}{120} f^{v}(\hat{x}) + \cdots \tag{7.10}$$

We multiply (7.9) by r^2 and subtract (7.10) and the resulting equation has no term involving h^2. After rearrangement, the equation has the form

$$f'(\hat{x}) = \frac{1}{2hr(r^2-1)}[r^3 f(\hat{x}+h) - f(\hat{x}+rh) - r^3 f(\hat{x}-h) + f(\hat{x}-rh)]$$

$$-\frac{r^2h^4}{120} f^{\mathrm{v}}(\hat{x}) + \cdots \qquad (7.11)$$

and we have obtained an expression for $f'(\hat{x})$ which is accurate to order h^4 rather than order h^2. This process can be repeated by combining yet more function values to remove more of the terms in the error expansion. In this way successive approximations with errors of order h^6, h^8 etc., can be obtained without the necessity to take smaller and smaller values of h.

One of the most common choices for r is $r=\frac{1}{2}$, in which case (7.11) takes the form

$$f'(\hat{x}) = \frac{-1}{6h}[f(\hat{x}+h) - 8f(\hat{x}+h/2) - f(\hat{x}-h) + 8f(\hat{x}-h/2)]$$

$$-\frac{h^4}{480} f^{\mathrm{v}}(\hat{x}) + \cdots$$

Using this formula with h=0.04 (the best value from Table 1) we obtain

$$f'(1.0000) = \frac{-1}{6(0.04)}[f(1.04) - 8f(1.02) - f(0.96) + 8f(0.98)]$$

$$= 2.7183$$

which is much more accurate.

The Richardson extrapolation technique has many applications other than to the field of numerical differentiation. In fact, we shall see it used again in the next section on numerical integration.

7.2 Integration

In a direct parallel with the situation in the differentiation section, numerical integration formulae are obtained by integrating polynomials which approximate the function in some way.

7.2.1 Newton–Cotes formulae

One obvious way to obtain an integration formula is to integrate an interpolation polynomial. For example, suppose that we can obtain a formula by integrating the *linear* Lagrangian interpolation polynomial. If we define $x_0 \equiv a$ and $x_1 \equiv b$, then we have

$$\int_a^b f(x)\,dx \equiv \int_{x_0}^{x_1} f(x)\,dx$$

$$\simeq \int_{x_0}^{x_1} \left[\frac{(x-x_1)}{(x_0-x_1)} f(x_0) + \frac{(x-x_0)}{(x_1-x_0)} f(x_1) \right] dx$$

$$= \frac{(x_1-x_0)}{2} [f(x_0) + f(x_1)]$$

If we let $h=b-a\equiv x_1-x_0$, then we can rewrite this as

$$\int_{x_0}^{x_1} f(x)\,dx \simeq \frac{h}{2} (f_0 + f_1) \tag{7.12}$$

where, as usual, $f_0\equiv f(x_0)$ and $f_1\equiv f(x_1)$. Formula (7.12) is known as the *trapezoidal* rule. The derivation of the name is obvious; the right-hand side is merely the area under the line joining f_0 and f_1; this is a trapezium whenever f_0 and f_1 have the same sign.

As we might expect, a more accurate formula can be obtained by splitting the interval $[a,b]$ into two equal subintervals of size $h=(b-a)/2$ and integrating the quadratic interpolation polynomial. If we define $x_0\equiv a$, $x_1=a+h$ and $x_2=a+2h=b$, then we have that

$$\int_a^b f(x)\,dx = \int_{x_0}^{x_2} f(x)\,dx$$

$$\simeq \int_{x_0}^{x_2} \left[\frac{(x-x_1)(x-x_2)}{(x_0-x_1)(x_0-x_2)} f(x_0) \right.$$

$$\left. + \frac{(x-x_0)(x-x_2)}{(x_1-x_0)(x_1-x_2)} f(x_1) + \frac{(x-x_0)(x-x_1)}{(x_2-x_0)(x_2-x_1)} f(x_2) \right] dx$$

$$= \frac{h}{3} [f_0 + 4f_1 + f_2]$$

This formula is the well-known *Simpson's* rule.

We can obtain formulae for the error resulting from the result of these numerical integration techniques. For example,

Theorem 7.1

The error associated with the trapezoidal approximation to the integral

$$\int_{x_0}^{x_1} f(x)\,dx$$

is

$$E_T = \frac{-h^3}{12} f''(\eta) \tag{7.13}$$

where $x_0<\eta<x_1$.

Proof

The error is the integral of the interpolation error, thus

$$E_T = \int_{x_0}^{x_1} \frac{(x-x_0)(x-x_1)}{2!} f''(\xi)\,dx \tag{7.14}$$

The integral cannot be evaluated directly because ξ is an unknown function of x. This problem can be circumvented by using the mean value theorem for integrals (Flett, 1966, p.230, Ex.6.8); the term $(x-x_0)(x-x_1)$ does *not* change sign in the interval $[x_0,x_1]$ and so (7.14) can be replaced by

$$E_T = \frac{f''(\eta)}{2!} \int_{x_0}^{x_1} (x-x_0)(x-x_1)\,dx$$

where η is some value with $x_0<\eta<x_1$. Performing the integration, we obtain

$$E_T = \frac{-h^3}{12} f''(\eta)$$

the required result. ■

Unfortunately we cannot obtain an error estimate for Simpson's rule in exactly the same way because the term $(x-x_0)(x-x_1)(x-x_2)$, which appears in the interpolation error formula, *does* change sign in the interval $[x_0,x_2]$.

However, using a different approach (see Johnson and Riess (1982), p.299), it is possible to show that the error incurred by using Simpson's rule is given by

$$E_S = \frac{-h^5}{90} f^{iv}(\eta) \qquad \text{for } x_0 < \eta < x_2 \tag{7.15}$$

If we examine the errors (7.13) and (7.15) closely, then an interesting result emerges. The equation (7.13) involves only the second derivative of the function $f(x)$ and so the trapezoidal error is zero if $f(x)$ is a linear polynomial; we would expect this because we have used linear interpolation to obtain the trapezoidal rule. However, (7.15) is written in terms of the *fourth* derivative of $f(x)$ resulting in Simpson's rule being exact for all polynomials of degree three, one higher than expected. This unexpected increase in order is particularly associated with the *Newton–Cotes* family of integration formulae, of which the trapezoidal and Simpson rules are members.

The Newton–Cotes formulae are obtained by subdividing the interval $[a,b]$ into n, say, equal subintervals of size $h=(b-a)/n$ and then integrating an interpolation polynomial defined on the points $x_i=a+ih$ ($x_0 \equiv a$, $x_n \equiv b$). The Newton–Cotes *closed* formulae integrate interpolation polynomials which use *all* the $(n+1)$ points; the trapezoidal and Simpson rules are closed formulae. Newon–Cotes *open* formulae do not involve all the data points. For example, if we divide $[a,b]$ into four equal subintervals using the points x_0 ($\equiv a$), x_1, x_2, x_3 and x_4($\equiv b$) and integrate the quadratic polynomial, defined by the function values at x_1, x_2 and x_3, over the whole interval, then the open formula, *Milne's* rule, is obtained. Thus

$$\int_{x_0}^{x_4} f(x)\,dx \simeq \frac{4h}{3}(2f_1 - f_2 + 2f_3) \tag{7.16}$$

with associated error

$$E_M = \frac{14h^5}{45} f^{iv}(\eta) \qquad \text{with } x_0 < \eta < x_4$$

Theorem 7.2

(a) for Newton–Cotes closed formulae
   (i) if n is odd then the error is of order h^{n+2}, and
   (ii) if n is even then the error is of order h^{n+3}

(b) for Newton–Cotes open formulae
   (i) if n is odd then the error is of order h^n, and
   (ii) if n is even then the error is of order h^{n+1}.

Proofs of these results can be found in Isaacson and Keller (1966).

The closed formulae are more accurate than the open ones and there would seem to be no reason ever to use open formulae. However, as we shall see in Chapter 8, these formulae have an important part to play in the solution of ordinary differential equations.

7.2.2 Composite rules

If we wish to increase the accuracy of our approximation to the integral, then it would appear that all we have to do is to split the range of integration into a large number of subintervals and integrate the interpolation polynomial defined on the nodes of this partition. However, as we know, using higher and higher degrees of interpolating polynomial is not a good idea. So we borrow another idea from Chapter 6; we use the basic idea behind cubic splines. The interval is split into a number of subintervals and then the function is integrated over each subinterval separately. For example, let us suppose that the interval $[a,b]$ is split into n equal subintervals and define the nodes of the partition as $x_i = x_0 + ih$ $(i=0,1,\ldots,n)$, where $h=(b-a)/n$ and $x_0 \equiv a$, $x_n \equiv b$. Now the integral can be rewritten as

$$\int_a^b f(x)\,dx = \int_{x_0}^{x_n} f(x)\,dx = \int_{x_0}^{x_1} f(x)\,dx + \int_{x_1}^{x_2} f(x)\,dx + \cdots + \int_{x_{n-1}}^{x_n} f(x)\,dx \tag{7.17}$$

and each of the integrals in (7.17) can be approximated using some numerical integration scheme. If we use the trapezoidal rule, then

$$\int_{x_i}^{x_{i+1}} f(x)\,dx \simeq \frac{h}{2}(f_i + f_{i+1})$$

and so

$$\int_a^b f(x)\,dx \simeq \frac{h}{2}(f_0 + f_1) + \frac{h}{2}(f_1 + f_2) + \cdots + \frac{h}{2}(f_{n-1} + f_n)$$

$$= \frac{h}{2}(f_0 + 2f_1 + 2f_2 + \cdots + 2f_{n-1} + f_n) \tag{7.18}$$

This is known as the *composite* form of the trapezoidal rule; its error is given by a sum of errors of the form of (7.13)

$$E_T = \frac{-h^3}{12}[f''(\eta_1) + f''(\eta_2) + \cdots + f''(\eta_n)]$$

where $x_0 < \eta_1 < x_1 < \eta_2 < \cdots < x_{n-1} < \eta_n < x_n$

If each $f''(\eta_i)$ is bounded in modulus by a value F, then this error expression can be bounded as follows

$$|E_T| \leq \frac{h^3}{12} nF$$

but $n=(b-a)/h$, and so

$$|E_T| \leq \frac{h^2}{12}(b-a)F$$

If, instead, we split the integration in the following way

$$\int_{x_0}^{x_n} f(x)\,dx = \int_{x_0}^{x_2} f(x)\,dx + \int_{x_2}^{x_4} f(x)\,dx + \cdots + \int_{x_{n-2}}^{x_n} f(x)\,dx$$

for some even n, then we can derive the composite form of Simpson's rule, replacing each of the integrals on the right-hand side using his formula, to give

$$\int_{x_0}^{x_n} f(x)\,dx \simeq \frac{h}{3}(f_0 + 4f_1 + 2f_2 + 4f_3 + 2f_4 + \cdots + 2f_{n-2} + 4f_{n-1} + f_n) \tag{7.19}$$

The error in using formula (7.19), by the same argument as for the composite trapezoidal rule and with similar assumptions, has the form

$$|E_S| \leq \frac{h^4}{180}(b-a)F$$

Composite forms of any of the Newton–Cotes formulae can be obtained easily.

7.2.3 Romberg integration

Romberg integration is a technique which allies Richardson extrapolation. described in Section 7.1.2, to the trapezoidal rule in such a way that a high accuracy approximation to an integral can easily be found. The composite form of the trapezoidal rule approximation to the integral

$$\int_a^b f(x)\,dx$$

is

$$\int_a^b f(x)\,\mathrm{d}x \simeq \frac{h}{2}\,(f_0 + 2f_1 + 2f_2 + \cdots + 2f_{m-1} + f_m) \tag{7.20}$$

in which $h=(b-a)/m$, $x_0 \equiv a$ and $x_m \equiv b$. We have seen that the error formula for this rule can be written as

$$E_T = \frac{-h^2}{12}\,(b-a)\,f''(\eta) \qquad \text{with } x_0 < \eta < x_m$$

Theorem 7.3

The error formula for the trapezoidal rule has the form

$$E_T = \frac{-h^2}{12}[f'(b) - f'(a)] + \frac{(b-a)}{720}\,h^4 f^{iv}(\xi) \qquad \text{with } x_0 < \xi < x_m \tag{7.21}$$

This result will turn out to be useful and a proof can be found in Ralston and Rabinowitz (1978, p.124).

In its simplest form, the Romberg method starts by evaluating approximations, of the form (7.20), to the integral for a sequence of decreasing values of h. We start with the value $h_0=b-a$ and successively halve the subintervals so that $h_1=h_0/2$, $h_2=h_1/2$ and in general $h_k=h_{k-1}/2=(b-a)/2^k$. If we denote the right-hand side of (7.20) by $S_{k,0}$ when h is replaced by h_k (and, as a result, m is 2^k), then

$$S_{0,0} = \frac{(b-a)}{2}[f(a) + f(b)] = \frac{h_0}{2}[f(a) + f(b)]$$

$$S_{1,0} = \frac{h_1}{2}[f(a) + 2f(a+h_1) + f(b)] = \tfrac{1}{2}\frac{h_0}{2}[f(a)+f(b)] + h_1 f(a+h_1)$$

$$= \tfrac{1}{2}S_{0,0} + h_1 f(a+h_1)$$

$$S_{2,0} = \frac{h_2}{2}[f(a)+2f(a+h_2)+2f(a+2h_2)+2f(a+3h_2)+f(b)]$$

$$= \tfrac{1}{2}\,\frac{h_1}{2}[f(a)+f(b)+2f(a+2h_2)] + h_2[f(a+h_2)+f(a+3h_2)]$$

$$= \tfrac{1}{2}S_{1,0} + h_2[f(a+h_2)+f(a+3h_2)]$$

And, in general we can show that

$$S_{k,0} = \tfrac{1}{2}S_{k-1,0} + h_k \sum_{i=1}^{2^{k-1}} f[a+(2i-1)h_k] \qquad \text{for } k=1,2,\ldots$$

We illustrate this process by approximating the integral

$$\int_0^1 \frac{1}{1+x}\,dx = 0.6931$$

Thus

$$S_{0,0} = \frac{(1-0)}{2}\left[\frac{1}{1+0}+\frac{1}{1+1}\right] = \frac{3}{4} = 0.75$$

$$S_{1,0} = \frac{0.75}{2} + 0.5\left[\frac{1}{1+0.5}\right] = 0.375 + 0.3333 = 0.7083$$

$$S_{2,0} = \frac{0.7083}{2} + 0.25\left[\frac{1}{1+0.25}+\frac{1}{1+0.75}\right]$$

$$= 0.35415 + 0.25[\,0.8+0.57143\,]$$

$$= 0.35415 + 0.34286 = 0.6970$$

correct to four decimal places. We can see that the approximations $S_{k,0}$ are converging to the correct answer, but slowly. We can speed up the process by applying the Richardson extrapolation technique. In general, using (7.21), we have

$$\int_a^b f(x)\,dx = S_{n-1,0} - \frac{h_{n-1}^2}{12}[f'(b) - f'(a)] + \frac{(b-a)}{720}h_{n-1}^4 f^{iv}(\xi_{n-1})$$

$$\text{where } a<\xi_{n-1}<b \qquad (7.22)$$

Similarly, for the step size $h_n = h_{n-1}/2$, we have that

$$\int_a^b f(x)\,dx = S_{n,0} - \frac{h_n^2}{12}[f'(b) - f'(a)] + \frac{(b-a)}{720} h_n^4 f^{iv}(\xi_n)$$

$$= S_{n,0} - \frac{h_{n-1}^2}{48}[f'(b) - f'(a)] + \frac{(b-a)}{720} h_n^4 f^{iv}(\xi_n)$$

$$\text{where } a<\xi_n<b \tag{7.23}$$

Combining (7.22) and (7.23) in such a way as to eliminate the terms in h_{n-1}^2, we obtain the equation

$$\int_a^b f(x)\,dx = \frac{4S_{n,0} - S_{n-1,0}}{3} + \frac{(b-a)}{2160}[4h_n^4 f^{iv}(\xi_n) - h_{n-1}^4 f^{iv}(\xi_{n-1})]$$

Thus

$$\int_a^b f(x)\,dx \simeq \frac{4S_{n,0} - S_{n-1,0}}{3}$$

with an error of order h_n^4 provided that f^{iv} is bounded. We now define

$$S_{k,1} = \frac{4S_{k,0} - S_{k-1,0}}{3} \quad \text{for each } k=1,2,\ldots \tag{7.24}$$

We can apply Richardson extrapolation to the approximations given by (7.24) to obtain approximations with successively higher powers of h in their error formulae. In general, we can show that

$$S_{k,j} = \frac{4^j S_{k,j-1} - S_{k-1,j-1}}{4^j - 1} \quad \text{for } k=1,2,\ldots, j=1,2,\ldots,k \tag{7.25}$$

It is interesting to note that the approximations produced using (7.25) are exactly those produced by successively higher-order Newton–Cotes formulae. For example, (7.24) is equivalent to Simpson's rule with $h=h_k$.

It is usual to present the approximations (7.25) in the form of a table:

$$\begin{array}{llll}
S_{0,0} & & & \\
S_{1,0} & S_{1,1} & & \\
S_{2,0} & S_{2,1} & S_{2,2} & \\
\cdot & \cdot & \cdot & \\
\cdot & \cdot & \cdot & \\
\cdot & \cdot & \cdot & \\
S_{n,0} & S_{n,1} & \ldots & S_{n,n}
\end{array}$$

Ralston and Rabinowitz (1978, p.125) prove that the diagonal elements, $S_{k,k}$, converge to the integral provided that the elements, $S_{k,0}$, in the first column do so. We would expect the diagonal elements to converge faster than those in the first column.

The most efficient use of this table results from performing the calculations row by row; then only one further application of the trapezoidal rule enables the next row to be calculated. Thus, the elements are calculated in the order $S_{0,0}$; $S_{1,0}$, $S_{1,1}$; $S_{2,0}$, $S_{2,1}$, $S_{2,2}$;. . . . When two diagonal elements $S_{k-1,k-1}$ and $S_{k,k}$ agree to the required tolerance, then the process stops and $S_{k,k}$ is accepted as the approximation to the integral.

Again we illustrate this process with reference to the integral

$$\int_0^1 \frac{1}{1+x}\,dx$$

We have already calculated $S_{0,0}$ and $S_{1,0}$ and so

$$S_{1,1} = \frac{4S_{1,0} - S_{0,0}}{3} = \frac{4(0.7083) - 0.75}{3} = 0.6044$$

Similarly, since we have also calculated $S_{2,0}$=0.6970, we have

$$S_{2,1} = \frac{4S_{2,0} - S_{1,0}}{3} = \frac{4(0.6970) - 0.7083}{3} = 0.6932$$

and now we calculate

$$S_{2,2} = \frac{4^2S_{2,1} - S_{1,1}}{4^2 - 1} = \frac{16S_{2,1} - S_{1,1}}{15} = \frac{16(0.6932) - 0.6944}{15}$$
$$= 0.6931$$

Thus, after these calculations, the table has the form

$S_{k,j}$	j=0	1	2
k=0	0.75		
1	0.7083	0.6944	
2	0.6970	0.6932	0.6931

and we can see that, if we require only two decimal place accuracy, $S_{2,2}$ and $S_{1,1}$ agree, whereas $S_{2,0}$ and $S_{1,0}$ do not. If two decimal place accuracy is not sufficient the next step is to calculate $S_{3,0}$ and then $S_{3,1}$, $S_{3,2}$ and, finally, $S_{3,3}$ which should be compared with $S_{2,2}$.

The procedure given in Fig. 7.1 implements Romberg integration. The process is terminated when two successive diagonal elements agree to within the required tolerance, or is abandoned if the number of rows required exceeds some predetermined limit. In the version presented, this limit is 10. Because each new value $S_{k,j}$ ($j>0$) requires only one value ($S_{k-1,j-1}$) from the previous row, only one row needs to be stored. The vector *s* stores the row. So that a new value $S_{k,j}$ can be stored in *s*[*j*], the value $S_{k-1,j}$ (currently stored in *s*[*j*])is stored temporarily (in *ss*) until $S_{k,j+1}$ has been computed.

```
procedure Romberg (function f (x : real) : real;
      a, b, tol : real;
      var integral : real;   var success : boolean);
  const
    maxrow = 10;   maxpt = 512;   {= 2^(maxrow-1)}
  type
    rows = 0 . . maxrow;
  var
    ss, news, sigma, hcoeff, h : real;
    twotothekminus1, fourtothej : 1 . . maxint;
    j, k : rows;
    p : 1 . . maxpt;
    s : array [rows] of real;
    state : (splitting, withintol, lastrowreached);
begin
  k := 1;   twotothekminus1 := 1;   h := (b-a)/2;
  news := h*(f(a)+f(b));   s[0] := news;
  state := splitting;
  repeat
    sigma := 0;   hcoeff:= -1;
    for p := 1 to twotothekminus1 do
    begin
      hcoeff := hcoeff + 2;
      sigma := sigma + f(a+hcoeff*h)
    end;
    ss := s[0];
    news := ss/2 + h*sigma;   s[0] := news;
    fourtothej := 1;
    for j := 1 to k - 1 do
    begin
      fourtothej := fourtothej * 4;
      news := (fourtothej*news-ss)/(fourtothej-1);
      ss := s[j];   s[j] := news
    end;
```

```
      fourtothej := fourtothej*4;
      news := (fourtothej*news-ss)/(fourtothej-1);
      if abs(news-ss) <= tol then state := withintol else
        if k = maxrow then state := lastrowreached else
        begin
          k := k + 1;   h := h/2;
          twotothekminus1 := twotothekminus1 * 2;
          s[k] := news
        end
    until state <> splitting;
    success := state = withintol;
    if success then integral := news
  end {Romberg};
```

Figure 7.1

7.2.4 Gaussian quadrature

The integration, or *quadrature*, rules that we have discussed have been restricted in the sense that only equally spaced quadrature points have been used. If we remove the restriction that the points be equally spaced, then we can design a more accurate formula based on a given number of points. For example, equations (7.13) and (7.15) show that the trapezoidal rule is *exact* for polynomials of degree one and similarly Simpson's rule is exact for polynomials of degree three. The trapezoidal rule approximation to

$$\int_{x_0}^{x_1} f(x)\,\mathrm{d}x$$

makes use of two function values, $f(x_0)$ and $f(x_1)$, evaluated at the ends of the integration interval, and two constants, both of which are $h/2=(x_1-x_0)/2$, multiplying the function values. If we remove the restriction that the points x_0 and x_1 be used for the quadrature rule, then we have four free parameters which we can use to design a new rule. With four parameters we can integrate a polynomial of degree *three* exactly and this is a significant increase in the precision of the integration rule. Similarly, using three points and three coefficients, the same number as for Simpson's rule, we can define a rule which is exact for polynomials of degree *five*, rather than *three*.

To introduce the ideas involved in Gaussian quadrature, we consider the more general integral

$$\int_{a}^{b} \omega(x)\, f(x)\,\mathrm{d}x$$

where $\omega(x)>0$ is a weight function. We are interested only in the case $\omega(x)=1$ but different choices do play very important roles in numerical integration and a discussion of these can be found in Ralston and Rabinowitz (1978).

The quadrature rules that we have discussed so far have been derived by replacing $f(x)$ by an interpolation polynomial $p_n(x)$, leading to the following approximation to the integral

$$I(f) = \int_a^b \omega(x)f(x)\,\mathrm{d}x \simeq \int_a^b \omega(x)p_n(x)\,\mathrm{d}x = \sum_{j=0}^{n} \alpha_j f_j = I_n(f) \qquad (7.26a)$$

where, for example, using Lagrangian interpolation,

$$\alpha_j = \int_a^b \omega(x)l_j(x)\,\mathrm{d}x \qquad (7.26b)$$

This quadrature rule is *exact* if $f(x)$ is a polynomial of degree at most n; the quadrature rule is thus said to have *precision n*. In particular the rule is exact if $f(x)$ is replaced by one of the powers x^k $(k=0,1,\ldots,n)$ and so

$$\int_a^b x^k\omega(x)\,\mathrm{d}x = \sum_{j=0}^{n} \alpha_j x_j^k \qquad \text{for } k = 0,1,\ldots,n$$

If we examine the right-hand side of this equation closely we note that there are $2n+2$ parameters involved: $n+1$ coefficients α_j and $n+1$ points x_j. We are tempted to ask if it is possible to construct a rule which, instead, will exactly integrate polynomials of degree $2n+1$. This is precisely the property possessed by Gaussian quadrature formulae.

To derive these formulae we must make use of orthogonal polynomials.

Definition 7.1

Let the polynomial $\phi_i(x)$ have degree i, then the set of polynomials $\{\phi_0(x),\phi_1(x),\ldots,\phi_{n+1}(x)\}$ is said to be *orthogonal* on the interval $[a,b]$, with respect to the weight function $\omega(x)>0$, if

$$\int_a^b \omega(x)\phi_i(x)\phi_j(x)\,\mathrm{d}x$$

is zero when $i\neq j$ and positive otherwise.

One of the main uses of orthogonal polynomials derives from the result of the following theorem.

Theorem 7.4

Let $q(x)$ be a polynomial of degree at most n; then $q(x)$ has the unique representation

$$q(x) = \sum_{j=0}^{n} \gamma_j \phi_j(x)$$

in terms of the orthogonal polynomials $\phi_j(x)$ ($j=0,1,\ldots,n$).

Proof

We assume that the representation is not unique; that is, there is another representation

$$q(x) = \sum_{j=0}^{n} \sigma_j \phi_j(x)$$

The difference

$$\sum_{j=0}^{n} (\gamma_j - \sigma_j)\phi_j(x)$$

must be identically zero and so, multiplying it by the arbitrary orthogonal polynomial, $\phi_k(x)$, the following identity must also hold:

$$\int_a^b \sum_{j=0}^{n} (\gamma_j - \sigma_j)\phi_j(x)\phi_k(x)\omega(x)\,dx = 0$$

But, using the orthogonality of the $\phi_j(x)$, this can be rewritten as

$$(\gamma_k - \sigma_k)\int_a^b \phi_k^2(x)\omega(x)\,dx = 0$$

Now, by definition, the integral cannot be zero and so

$$\gamma_k = \sigma_k$$

and, because the index k was arbitrary, the representation is unique.

Johnson and Riess (1981, p.325) prove the following theorem.

Theorem 7.5

Let $\{\phi_j(x)\}$ $(j=0,1,\ldots,n+1)$ be a set of orthogonal polynomials as given above. Then for each $j \geq 0$, the zeros of $\phi_j(x)$ are real and distinct and lie in the interval (a,b).

The enhanced precision of Gaussian quadrature rules is illustrated by the following theorem.

Theorem 7.6

If $p_{2n+1}(x)$ is a polynomial of degree $2n+1$, then the formula

$$\int_a^b p_{2n+1}(x)\omega(x)\,dx = \sum_{j=0}^{n} \alpha_j p_{2n+1}(x_j)$$

is exact if the points x_j $(j=0,1,\ldots,n)$ are the zeros of the orthogonal polynomial $\phi_{n+1}(x)$ and the α_j are as given in (7.26b).

Proof

The polynomial $p_{2n+1}(x)$ can be written in the form

$$p_{2n+1}(x) = q(x)\phi_{n+1}(x) + r(x)$$

where $q(x)$ and $r(x)$ are polynomials of degree at most n. Now, because the x_j $(j=0,1,\ldots,n)$ are the zeros of $\phi_{n+1}(x)$, it follows that

$$p_{2n+1}(x_j) = r(x_j) \qquad \text{for } j = 0,1,\ldots,n$$

Also, because the quadrature rule (7.26a) has precision n,

$$\int_a^b r(x)\omega(x)\,dx = \sum_{j=0}^{n} \alpha_j r(x_j)$$

Now consider

$$I(p_{2n+1}) = \int_a^b p_{2n+1}(x)\omega(x)\,dx$$

$$= \int_a^b [q(x)\phi_{n+1}(x) + r(x)]\omega(x)\,dx$$

$$= \int_a^b q(x)\phi_{n+1}(x)\omega(x)\,dx + \int_a^b r(x)\omega(x)\,dx$$

Using the result of Theorem 7.4, $q(x)$ is replaced by its unique representation

$$q(x) = \sum_{j=0}^{n} \gamma_j \phi_j(x)$$

and the first term in $I(p_{2n+1})$ is then automatically zero because $\phi_{n+1}(x)$ is orthogonal to all the other $\phi_j(x)$ $(j=0,1,\ldots,n)$. Thus

$$I(p_{2n+1}) = \int_a^b r(x)\omega(x)\,dx = \sum_{j=0}^{n} \alpha_j r(x_j)$$

$$= \sum_{j=0}^{n} \alpha_j p_{2n+1}(x_j) = I_n(p_{2n+1})$$

and the theorem is proved. ■

The following theorem, the converse of Theorem 7.6, is proved by Johnson and Riess (1982, p.326).

Theorem 7.7

The quadrature formula given above has precision $2n+1$ only if the x_j $(j=0,1,\ldots,n)$ are the roots of $\phi_{n+1}(x)$.

They also prove the important result that the weights α_j $(j=0,1,\ldots,n)$ are positive and so the quadrature formula has good properties for the control of rounding errors.

The following estimate of the error incurred in this type of quadrature is derived in Ralston and Rabinowitz (1978, p.103). The appearance of $f^{(2n+2)}$ in the error expression again verifies the precision of the quadrature rule.

Theorem 7.8

If $f(x)$ has $2n+2$ continuous derivatives in (a,b), then the error in using Gaussian quadrature is

$$I(f) - I_n(f) = \frac{f^{(2n+2)}(\eta)}{(2n+2)!} \int_a^b q_{n+1}^2(x)\omega(x)\,dx$$

where $q(x)$ is the *monic Chebyshev* polynomial of the *first* kind.

In our case the weight function is $\omega(x)=1$ and the orthogonal polynomials corresponding to this weight function are known as the *Legendre* polynomials and are denoted by $P_j(x)$ ($j=0,1,\ldots$). Quadrature using these polynomials is called Gauss–Legendre quadrature or, simply, Gaussian quadrature.

We can calculate the coefficients α_j ($j=0,1,\ldots,n$) but this is not necessary because they, and the points x_j, have already been tabulated for a large range of values of n {See Stroud and Secrest (1966)}. Some of the roots of the Legendre polynomials and the corresponding weights are given in Table 7.2.

Table 7.2

n	*Roots* x_j	*Weights* α_j
1	±0.577550	1
2	0	0.888889
	±0.774597	0.555556
3	±0.339981	0.652145
	±0.861136	0.347855

We need not worry about the fact that the derivation given above is dependent on the integral being defined on the interval $[-1,1]$ since any interval $[a,b]$ can be transformed into $[-1,1]$ by using the simple linear transformation

$$y = \frac{2x - a - b}{b - a}$$

and so

$$\int_a^b f(x)\,\mathrm{d}x = \tfrac{1}{2}\int_{-1}^{1} f\left(\frac{(b-a)y+b+a}{2}\right)(b-a)\,\mathrm{d}y$$

The increase in accuracy gained by using Gaussian quadrature rules is easily demonstrated by considering an example; we evaluate

$$\int_{-1}^{1} \mathrm{e}^x\,\mathrm{d}x = \mathrm{e}^1 - \mathrm{e}^{-1} \simeq 2.3504$$

numerically. The trapezoidal rule gives the approximation

$$\int_{-1}^{1} \mathrm{e}^x\,\mathrm{d}x \simeq \tfrac{2}{2}(\mathrm{e}^1 + \mathrm{e}^{-1}) = 3.0862$$

The two-point Gaussian rule uses the points -0.57755 and 0.57755 and coefficients 1 and 1. This gives the result

$$\int_{-1}^{1} \mathrm{e}^x\,\mathrm{d}x \simeq 1(\mathrm{e}^{-0.57755} + \mathrm{e}^{0.57755}) = 2.3429$$

which is very much more accurate.

7.2.5 Adaptive quadrature

All the rules discussed so far suffer from the disadvantage that they assume that the function behaves uniformly; this is unreasonable because most functions vary more rapidly in one part of the interval than in another. An *adaptive* quadrature formula can take this local variation into account. We illustrate the technique of adaptive quadrature by using the trapezoidal rule to approximate

$$I = \int_a^b f(x)\,\mathrm{d}x$$

in such a way that S_{ap}, the approximation to the integral, is accurate to within a specified tolerance ϵ; that is

$$|I - S_{ap}| \leq \epsilon$$

To achieve this the interval $[a,b]$ is split up into a sequence of subintervals $[x_{i-1},x_i]$ of length $h_{i-1}=x_i-x_{i-1}$; from (7.13) the contribution to the total error from each such subinterval is

$$-\frac{h_{i-1}^3}{12}f''(\eta_i) \qquad x_{i-1}<\eta_i<x_i$$

The size of this error depends on h_{i-1} and $f''(\eta_i)$; when $|f''(\eta_i)|$ is 'large', then a smaller h_{i-1} is necessary to control the error. Thus the sizes of the subintervals are chosen so that the contribution from each subinterval to the total error is approximately equal. If $I(h_{i-1})$ and $S_{ap}(h_{i-1})$ are respectively the integral and its approximation on the interval $[x_{i-1},x_i]$, and if

$$|I(h_{i-1}) - S_{ap}(h_{i-1})| \leq \frac{h_{i-1}\epsilon}{(b-a)}$$

then the total error is kept below ϵ.

The process is started by forming the trapezoidal approximation on the whole interval $[a,b]$. Thus

$$S_0^{(1)} = \frac{(b-a)}{2}[f(a)+f(b)] = \frac{h_0}{2}[f(a)+f(b)]$$

Now a second approximation, $S_0^{(2)}$, is obtained by splitting $[a,b]$ into two equal subintervals of size $h_1=h_0/2$. Thus

$$\begin{aligned} S_0^{(2)} &= \frac{1}{2}\,\frac{(b-a)}{2}[f(a) + 2f(a+h_0/2) + f(b)] \\ &= \tfrac{1}{2}[S_0^{(1)} + h_0 f(a+h_0/2)] \\ &= \tfrac{1}{2}S_0^{(1)} + h_1 f(a+h_1) \end{aligned}$$

From (7.13) we can write down the errors in these two approximations

$$I_0 - S_0^{(1)} = \frac{-f''(\eta_1)}{12}h_0^3 \qquad a<\eta_1<b \tag{7.27}$$

and

$$I_0 - S_0^{(2)} = \frac{-2f''(\eta_2)}{12}\left(\frac{h_0}{2}\right)^3 \qquad a<\eta_2<b \tag{7.28}$$

In (7.28), the factor 2 occurs because we are integrating over two adjacent subintervals. Assuming that $f''(x)$ is approximately constant on the interval $[a,b]$ (and denoted by $f''(\eta)$), we can subtract (7.28) from (7.27) and obtain

$$S_0^{(2)} - S_0^{(1)} \simeq \frac{f''(\eta)}{12}\left(\frac{h_0}{2}\right)^3(2-2^3)$$

from which

$$\frac{f''(\eta)}{12}\left(\frac{h_0}{2}\right)^3 \simeq \frac{S_0^{(2)} - S_0^{(1)}}{-6} \tag{7.29}$$

Substituting (7.29) into the right-hand side of (7.28) we obtain the error estimate

$$I_0 - S_0^{(2)} \simeq \tfrac{1}{3}[S_0^{(2)} - S_0^{(1)}] \tag{7.30}$$

Thus, the error in the more accurate approximation, $S_0^{(2)}$, is approximately one third of the difference $S_0^{(2)} - S_0^{(1)}$ and we can easily calculate this.

If the error

$$E_0 = \tfrac{1}{3} \mid S_0^{(2)} - S_0^{(1)} \mid \leq \epsilon \tag{7.31}$$

then we will accept $S_0^{(2)}$ as our approximation to I_0. If (7.31) is not satisfied, then we restrict our attention first to the subinterval $[a,a+h_1]$ and then to the subinterval $[a+h_1,b]$ and attempt to obtain approximations to the integrals in those subintervals which are accurate, in the sense of (7.31), to within a tolerance of $\epsilon/2$. This process is repeated until the approximation to the integral over each subinterval is sufficiently accurate.

We illustrate this adaptive process by considering

$$\int_0^1 \frac{1}{1+x}\,dx = [\ln(1+x)]_0^1 = \ln 2 \simeq 0.6931$$

which we wish to evaluate numerically, accurate to within $\epsilon=0.005$. Firstly we evaluate $S_0^{(1)}$, an approximation to the integral over the whole range of integration

$$S_0^{(1)} = \frac{1}{2}\left(\frac{1}{1}+\frac{1}{2}\right) = \frac{3}{4} = 0.75$$

Now, splitting the interval into two we obtain

$$S_0^{(2)} = \frac{0.5}{2}\left[\frac{1}{1} + 2\left(\frac{1}{1+0.5}\right) + \frac{1}{2}\right] = 0.7083$$

and

$$\frac{1}{3} \mid S_0^{(2)} - S_0^{(1)} \mid = 0.0139 \not< 0.005 = \epsilon$$

Thus, the error test is *not* satisfied for this interval size and so we concentrate on approximating the integral on the first interval, [0,0.5]. In this case $h=0.5$ and the error tolerance is $\epsilon/2=0.0025$. The first approximation to the integral on [0,0.5] is

$$S_0^{(1)} = \frac{0.5}{2}(1 + \tfrac{2}{3}) = 0.4167$$

and the second is

$$S_0^{(2)} = \frac{0.5}{4}[1 + 2(\tfrac{4}{5}) + \tfrac{2}{3}] = 0.4083$$

The error estimate is

$$\tfrac{1}{3} \mid S_0^{(2)} - S_0^{(1)} \mid = \tfrac{1}{3} \mid 0.0084 \mid = 0.0028 \not< 0.0025$$

and so the estimate, $S_0^{(2)}$, is *not* sufficiently accurate. Thus we split [0,0.5] into the two further subintervals [0,0.25] and [0.25,0.5] and consider the integral on each in turn. On [0,0.25] the error bound is $\epsilon/4=0.00125$ and

$$S_0^{(1)} = \frac{0.25}{2}(1 + \tfrac{4}{5}) = 0.2250$$

$$S_0^{(2)} = \frac{0.25}{4}[1 + 2(\tfrac{8}{9}) + \tfrac{4}{5}] = 0.2236$$

The error estimate is

$$\tfrac{1}{3} \mid S_0^{(2)} - S_0^{(1)} \mid = \tfrac{1}{3}(0.0014) = 0.0005 < 0.00125$$

and we have our first contribution to the integral over the whole interval. We will accumulate this total integral in the variable *sum* and so, to begin, *sum*=0.2236. Now we consider the interval [0.25,0.5] in which

$$S_1^{(1)} = \frac{0.25}{2}(\tfrac{4}{5} + \tfrac{2}{3}) = 0.1833$$

$$S_1^{(2)} = \frac{0.25}{4}[\tfrac{4}{5} + 2(\tfrac{8}{11}) + \tfrac{2}{3}] = 0.1826$$

The error estimate is

$$\tfrac{1}{3} \mid S_1^{(2)} - S_1^{(1)} \mid = \tfrac{1}{3}(0.0007) = 0.0002 < 0.00125$$

and so we accumulate a further contribution to the total

$$sum = sum + 0.1826 = 0.4062$$

We now turn our attention to the interval [0.5,1] on which the error bound, as for [0,0.5], is 0.0025. Here

$$S_2^{(1)} = \frac{0.5}{2}(\tfrac{2}{3} + \tfrac{1}{2}) = 0.2917$$

$$S_2^{(2)} = \frac{0.5}{4}(\tfrac{2}{3} + 2(\tfrac{4}{7}) + \tfrac{1}{2}) = 0.2887$$

and

$$\tfrac{1}{3} \mid S_2^{(2)} - S_2^{(1)} \mid = \tfrac{1}{3}(0.0030) = 0.0010 < 0.0025$$

and so the final estimate of the integral is

$$sum = sum + 0.2887 = 0.4062 + 0.2887 = 0.6949$$

We can see, immediately, that the actual error

$$\mid 0.6931 - 0.6949 \mid = 0.0018 < 0.005$$

and so the use of the adaptive quadrature method has been successful.

Adaptive quadrature rules can be based on *any* numerical integration scheme; because of its accuracy (related to the number of points involved) adaptive rules are often based on Simpson's rule. The major difficulty associated with adaptive quadrature methods lies in the task of keeping track of the stage that has been reached; we must, for example, know over which subinterval we are integrating and which is the next subinterval that we must consider. Obviously, if the derivatives of f change rapidly in some interval then we may need to create a large number of nested subintervals in order to calculate the integral to the required accuracy.

Fortunately there is no problem when our programming language allows recursion (see Section 1.5.4). The function of Fig. 7.2 uses the trapezoidal rule of (7.12) as the basis for an adaptive quadrature method. It evaluates the integral on the interval $[x_0,x_1]$ assuming convergence will be achieved.

```
function int (x0, x1, errtol : real) : real;
  var
    h, s1, s2 : real;
begin
  h := x1 − x0;
  s1 := (f(x0) + f(x1)) * h/2;
  s2 := (s1 + h * f(x0+h/2)) / 2;
  if abs(s1 − s2)/3 <= errtol then int := s2 else
    int := int (x0, x0+h/2, errtol/2) + int (x0+h/2, x1, errtol/2)
end {int};
```

Figure 7.2

There are two simple things we can do to improve efficiency. First, we can reduce the number of computations of $x0+h/2$ by storing this value in a local variable. Second, we can reduce the number of function evaluations by taking the function values computed at one recursive level and passing them as parameters to the next. Both are done in Fig. 7.3.

```
function int (x0, f0, x1, f1, errtol : real) : real;
  var
    h, s1, s2, xmid, fmid : real;
begin
  h := x1 − x0;   xmid := x0 + h/2;   fmid := f(xmid);
  s1 := (f0+f1) * h/2;   s2 := (s1 + h * fmid)/2;
  if abs(s1 − s2)/3 <= errtol then int := s2 else
  int := int (x0, f0, xmid, fmid, errtol/2) +
         int (xmid, fmid, x1, f1, errtol/2)
end {int};
```

Figure 7.3

To present a convenient interface to the user, this function can be incorporated within another.

```
function integral (function f (x : real) : real;
                  a, b, epsilon : real) : real;
  function int (x0, f0, x1, f1, errtol : real) : real;
    . . .
begin {integral}
  integral := int (a, f(a), b, f(b), epsilon)
end {integral};
```

There is one further test we must make. The function *int* is infinitely recursive if convergence is never achieved and so we must impose a minimum bound upon the size of interval considered. If an attempt is made to use an interval smaller than this minimum, the process must be abandoned immediately.

```
procedure TrapAdapt (function f (x : real) : real;
     a, b, epsilon, minh : real;
     var integral, lb, ub : real;    var success : boolean);
  procedure Interval
        (x0, f0, x1, f1, errtol : real;
         var int, lb, ub : real;    var ok : boolean);
     var
        h, xmid, fmid, s1, s2, int1, int2 : real;
  begin
     h := x1-x0;
     if h >= minh then
     begin
        xmid := x0 + h/2;   fmid := f(xmid);
        s1 := (f0 + f1) * h/2;   s2 := (s1 + h * fmid) / 2;
        if abs(s1 - s2)/3 <= errtol then int := s2 else
        begin
           Interval (x0, f0, xmid, fmid, errtol/2, int1, lb, ub, ok);
           if ok then
           begin
              Interval (xmid, fmid, x1, f1, errtol/2, int2, lb, ub, ok);
              if ok then
                 int := int1 + int2
           end
        end
     end else
     begin
        ok := false;   lb := x0;   ub := x1
     end
  end {Interval};


begin {Trap adapt}
  success := true;
  Interval (a, f(a), b, f(b), epsilon, integral, lb, ub, success)
end {Trap adapt};
```

Figure 7.4

```
procedure TrapAdapt (function f (x : real) : real;
    a, b, epsilon, minh : real;
    var integral, lb, ub : real;   var success : boolean);


  procedure Interval
      (x0, f0, x1, f1, errtol : real;
       var sum, lb, ub : real;   var ok : boolean);
    var
      h, xmid, fmid, s1, s2 : real;
      state : (chopping, withintol, htoosmall);
  begin
    state := chopping;
    repeat
      h := x1 - x0;
      if h < minh then state := htoosmall else
      begin
        xmid := x0 + h/2;   fmid := f(xmid);
        s1 := (f0+f1) * h/2;   s2 := (s1 + h * fmid)/2;
        if abs(s1 - s2)/3 <= errtol then state := withintol else
        begin
          errtol := errtol/2;
          Interval (x0, f0, xmid, fmid, errtol, sum, lb, ub, ok);
          x0 := xmid;   f0 := fmid
        end
      end
    until (state <> chopping) or not ok;

    if ok then
      case state of
        withintol : sum := sum + s2;
       htoosmall : begin
                     ok := false;   lb := x0;   ub := x1
                   end
      end {case}
  end { Interval };


begin {Trap adapt}
  success := true;   integral := 0;
  Interval (a, f(a), b, f(b), epsilon, integral, lb, ub, success)
end { Trap adapt };
```

Figure 7.5

Following our usual practice in such circumstances, we introduce a boolean variable parameter and, because the routine *integral* can now return more than one value, we write it as a procedure rather than as a function. The function *int* must now test the size of the interval and, if necessary, transfer control directly to the end of the outer routine. A function which, under some circumstances, could transfer control elsewhere and deliver no value would be displaying an extreme side effect and it was stated in Section 1.5 that functions with side effects are undesirable. Consequently this routine too is recast as a procedure. The final form is in Fig. 7.4. If the process breaks down because an interval becomes too small, the lower and upper bounds of that interval are returned via the variable parameters *lb* and *ub*.

Recursion is a sensible tool to use in this context but does carry some overhead because of the repeated procedure activations. In the present case this overhead is amplified because the procedure *Interval*, calls itself twice. We can replace one recursive call by a loop and this halves the number of procedure calls. In Fig. 7.5 a loop is used to accumulate the values resulting from integrating the second half of each interval and recursion is used only for the first half of each interval. The role of the final parameter of *Interval* has been changed; rather than compute the integral over the current interval it now updates a supplied variable by the integral over the current interval and so serves to accumulate the total integral.

Of course, some programming languages (Fortran, for example) do not permit recursion and then we must develop our own mechanism for keeping track of intervals yet to be integrated. The simplest technique is to use a *stack*, a data structure to which items can be added and from which they can subsequently be removed in the reverse order to that in which they were added.

If pointers are available (as in Pascal) no bound need be placed upon the maximum extent of the stack; otherwise a stack of fixed maximum size can be represented using an array. To illustrate the process of implementing adaptive quadrature without recursion, we shall assume the existence of a suitable data type *stacks* to represent a stack of cells each able to store a real value. We shall also assume the existence of two procedures *Save* and *Retrieve* to act upon this stack, the former adding five specified values to the stack and the latter removing five values from the stack and assigning them to five specified variables. These values will be the two end-points of an interval, the function values at these points and the error tolerance for the interval.

The algorithm now has the form shown in Fig. 7.6. The process is similar to that of Fig. 7.5 in that a loop is used to accumulate the integral over one half of each interval (in this case, the first half of each interval). Rather than use recursion to integrate over the second half of an interval, the relevant information for that half is saved and then retrieved later when the total integral for the first half has been computed. The outline of Fig. 7.6 does not check that the value of *maxcell* is sufficiently large for the stack to accommodate all the information saved. Because of the overhead of maintaining an explicit stack, the version of Fig. 7.5 may well be more efficient.

```
procedure TrapAdapt (function f : real;
    a, b, epsilon, minh : real;
    var integral, lb, ub : real;   var success : boolean);

  const
    bottom = 0;   maxcell = . . .;

    type
      cells = bottom . . maxcell;
      stacks = array [cells] of real;

    var
      stack : stacks;
      sp : cells;   {acts as stack pointer}
      h, xmid, fmid, s1, s2, x0, f0, x1, f1, sum : real;
      state : (chopping, htoosmall, done);

  procedure Save
      (var s : stacks;   var p : cells;   a, fa, b, fb, eps : real);
    . . .

  procedure Retrieve (var p : cells;   var a, fa, b, fb, eps : real);
    . . .

begin {Trap adapt}
  sp := bottom;
  x0 := a;   f0 := f(a);   x1 := b;   f1 := f(b);   sum := 0;
  state := chopping;
  repeat
    h := x1-x0;
    if h < minh then state := htoosmall else
    begin
      xmid := x0 + h/2;   fmid := f(xmid);
      s1 := (f0+f1) * h/2;   s2 := (s1+h * fmid)/2;
      if abs(s1 - s2)/3 <= epsilon then
      begin
        sum := sum + s2;
        if sp = bottom then state := done else
          Retrieve (sp, x0, f0, x1, f1, epsilon)
      end else
      begin
        epsilon := epsilon/2;
        Save (stack, sp, xmid, fmid, x1, f1, epsilon);
        x1 := xmid;   f1 := fmid
      end
    end
  until state <> chopping;
```

```
    success := state = done;
    case state of
      done : integral := sum;
      htoosmall : begin
                    lb := x0;   ub := x1
                  end
    end {case}
  end {Trap adapt};
```

Figure 7.6

7.3 Exercises

1 Modify the procedure *CubicSpline* of Fig. 6.4 so that it will return an estimate of the first or second derivative of the function at a given point. Apply this procedure to evaluate the first and second derivative of the function e^x at the point $x=1$, given the data

x	0.0	0.5	1.5
$f(x)$	1.0000	1.6487	4.4817

2 Write procedures employing the composite trapezoidal and Simpson's rules to estimate the integral of a function $f(x)$ given the equally spaced data (x_0,f_0), (x_1,f_1), . . ., (x_n,f_n). The Simpson procedure should check that n is *even*.

3 Apply both the procedures of Exercise 2 to approximate

$$\int_0^{\frac{\pi}{2}} \sin(x)\,dx = 1$$

using the data $(ih, \sin(ih))$ $(i=0,1,\ldots,20)$, where $h=\pi/40$.

4 Apply the procedure *Romberg* of Fig. 7.1 to the data of Exercise 3 using a tolerance of 0.00005.

5 Estimate

$$\int_0^{\frac{\pi}{2}} \sin(x)\,dx$$

by using the Gauss two-point rule on each of 10 equal subintervals. Compare your answers with those of Example 3.

6 Rewrite the function *int* of Fig. 7.3 and the procedure *TrapAdapt* of Fig. 7.4 to use Simpson's rule as the basic quadrature formula and apply the modified routines to the problem of evaluating

$$\int_0^{\frac{\pi}{2}} \sin(x)\,dx$$

correct to six decimal places.

7 Complete the procedure *TrapAdapt* of Fig. 7.6 by providing the two procedures *Save* and *Retrieve*. Test your procedure by applying it to the data of Exercise 3 and, if your Pascal implementation allows you to time sections of program, compare the run-time of your procedure with that of Fig. 7.5.

8 Apply the procedure *Romberg* and *TrapAdapt*, with varying ϵ and *minh*, in an attempt to evaluate the integral

$$\int_0^2 \frac{1}{1-x}\,dx$$

Both procedures should fail to perform the integration because of the singularity at $x=1$.

Chapter 8 ORDINARY DIFFERENTIAL EQUATIONS

We do not have to search very deeply into the vast store of mathematical models of the physical world to encounter ordinary differential equations; wherever there occurs a continuously changing process then a differential equation model is usually appropriate. In certain, relatively infrequent, cases we can find an analytic solution to the differential equation. However with most realistic models the differential equations are so complicated that only numerical methods offer a chance of determining a solution.

A whole book could be devoted to solving ordinary differential equations; in this chapter we shall give only a brief outline of some of the techniques used. There are, of course, many books available and the work appearing here is closely based on that appearing in Lambert (1973).

8.1 Theoretical aspects

Definition 8.1

In the differential equation

$$\frac{d^n y}{dx^n} = y^{(n)} = f(x,y,y',\ldots,y^{(n-1)})$$

the order of the highest derivative to appear is n; the differential equation is therefore said to be of nth *order*.

We shall discuss, almost exclusively, solving an *initial-value problem* for a first-order differential equation. A first-order differential equation $y'=f(x,y)$ may have an infinite family of solutions; for example, the differential equation

$$y' = \lambda y$$

where λ is a given constant, has the family of solutions

$$y = Ce^{\lambda x}$$

for any constant C. The solutions can be thought of as a family of curves, each curve being associated with a unique value of C. To isolate a particular curve we need to prescribe the value of C for that curve and this is achieved by imposing an *initial condition*, $y(a)=\eta$ (say), on the solution. We can verify that the solution satisfying this condition is

$$y = \eta e^{\lambda(x-a)}$$

This situation is illustrated in Fig. 8.1, in which a particular member of a family of curves has been selected by imposing an initial condition at the point $x=a$.

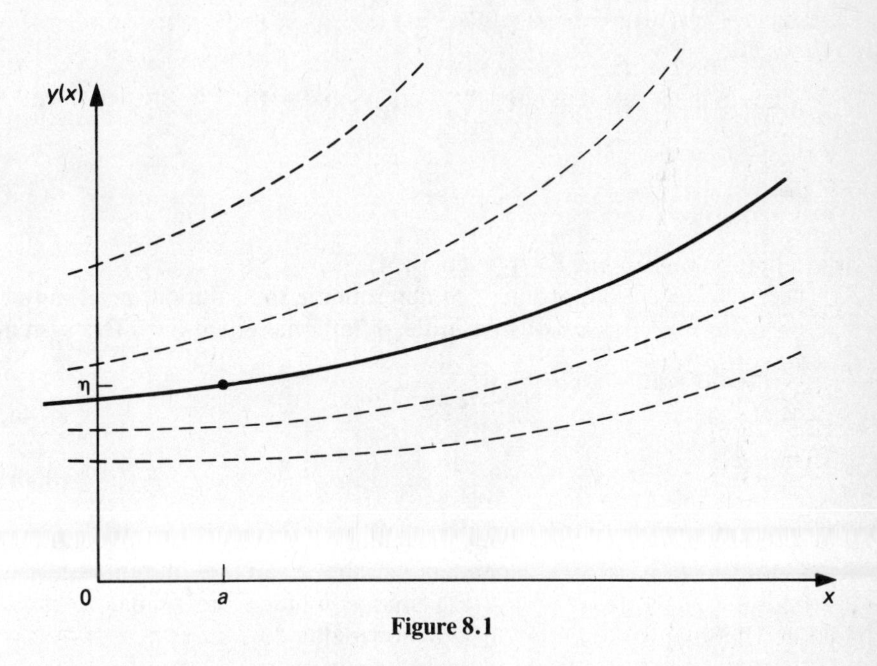

Figure 8.1

We say that the *pair* of equations

$$y' = f(x,y) \tag{8.1a}$$

$$y(a) = \eta \tag{8.1b}$$

constitutes an *initial-value problem*. The following theorem, the proof of which can be found in Henrici (1962), states a *sufficient* condition for a solution to (8.1) to exist and be unique.

Theorem 8.1

Let $f(x,y)$ be defined and continuous for all points (x,y) in a region D defined by $a \le x \le b$, $-\infty < y < \infty$ with a and b finite, and let there exist a constant L such that, for every (x,y_1) and (x,y_2) in D

$$|f(x,y_1) - f(x,y_2)| \leq L\,|y_1 - y_2| \tag{8.2}$$

Then there exists a unique solution $y(x)$ of the given initial-value problem (8.1), where $y(x)$ is continuous and differentiable for all (x,y) in D.

The inequality (8.2) is known as a *Lipschitz* condition for f, and L is known as a Lipschitz constant. If $f(x,y)$ is continuously differentiable with respect to y for all (x,y) in D, then f satisfies condition (8.2); using the mean value theorem (Flett 1966, p.140) there exists a value $\hat{y}$ such that

$$f(x,y_1) - f(x,y_2) = \frac{\partial}{\partial y}[f(x,\hat{y})](y_1 - y_2)$$

where $\hat{y}$ lies in the interval defined by y_1 and y_2 and so (8.2) is satisfied if we take

$$L = \sup_{(x,y)\text{ in }D} \left| \frac{\partial}{\partial y} f(x,y) \right| \tag{8.3}$$

For a definition of 'sup' see Flett (1966, p.74).

There also exists the problem of determining the solution of an initial-value problem for a *system* of first-order differential equations. The system takes the form

$$\mathbf{y}' = \mathbf{f}(x,\mathbf{y}) \tag{8.4a}$$

$$\mathbf{y}(a) = \boldsymbol{\eta} \tag{8.4b}$$

where $\mathbf{y}$ is the n-vector of unknowns, the initial-value $\boldsymbol{\eta}$ is also an n-vector and $\mathbf{f}$ is a vector function of the independent variable x and the solution vector $\mathbf{y}$. The conditions necessary to ensure a unique solution are similar to those stated in Theorem 8.1; the region D is now defined by $a \leq x \leq b$, $-\infty < y_i < \infty$ $(i=1,2,\ldots,n)$ and (8.2) is replaced by the condition

$$\|\mathbf{f}(x,\mathbf{y}_1) - \mathbf{f}(x,\mathbf{y}_2)\| \leq L\,\|\mathbf{y}_1 - \mathbf{y}_2\| \tag{8.5}$$

where $(x,\mathbf{y}_1)$ and $(x,\mathbf{y}_2)$ are in D and $\|\cdot\|$ denotes a vector norm. When $\mathbf{f}(x,\mathbf{y})$ is continuously differentiable with respect to *each* y_i, then we can replace (8.3) by

$$L = \sup_{(x,\mathbf{y})\text{ in }D} \left\| \frac{\partial \mathbf{f}}{\partial \mathbf{y}} \right\| \tag{8.6}$$

where $\partial\mathbf{f}/\partial\mathbf{y}$ is the *Jacobian* of $\mathbf{f}$ with respect to $\mathbf{y}$; i.e.

$$\left[\frac{\partial \mathbf{f}}{\partial \mathbf{y}}\right]_{i,j} = \frac{\partial f_i}{\partial y_j}$$

and $\| \cdot \|$, in (8.6), denotes a matrix norm, subordinate to the vector norm of (8.5). We shall not discuss the solution of systems of ordinary differential equations but, instead, refer the reader to Burden *et al.* (1981).

Ordinary differential equations also appear in *boundary-value problems.* In this case the differential equation must be at least second order and the conditions, necessary to identify a unique solution, must occur at different points. The following pair of equations

$$y'' = f(x,y,y')$$

$$y(a) = \eta \qquad y(b) = \mu$$

constitutes a boundary-value problem. Boundary-value problems also constitute an *extremely* important subset of problems governed by *partial* differential equations. Determining the existence and uniqueness of the solution to a boundary-value problem is more difficult than for an initial-value problem and we shall not discuss this at all. Instead, we refer the reader to Burden *et al.* (1981).

8.1.1 Taylor's series methods

One analytic method that we can use to attempt to determine a solution to the initial-value problem (8.1) is to use the Taylor's series expansion

$$y(x+h) = y(x) + h\frac{\mathrm{d}}{\mathrm{d}x}[y(x)] + \frac{h^2}{2!}\frac{\mathrm{d}^2}{\mathrm{d}x^2}[y(x)] + \frac{h^3}{3!}\frac{\mathrm{d}^3}{\mathrm{d}x^3}[y(x)] + \cdots \tag{8.7}$$

of $y(x+h)$ about the point x; we have written (8.7) in this form to emphasize that *total* derivatives with respect to x are used. Now, using the differential equation, we have that

$$\frac{\mathrm{d}^2}{\mathrm{d}x^2}[y(x)] = \frac{\mathrm{d}}{\mathrm{d}x}[f(x,y)] = \frac{\partial}{\partial x}[f(x,y)] + \frac{\partial}{\partial y}[f(x,y)]. \frac{\mathrm{d}y}{\mathrm{d}x} \tag{8.8}$$

and, using the notation $f = f(x,y)$ and $f_z = \partial f / \partial z$, this gives

$$\frac{\mathrm{d}^2}{\mathrm{d}x^2}[y(x)] = f_x + ff_y \tag{8.9}$$

Similarly

$$\frac{\mathrm{d}^3}{\mathrm{d}x^3}[y(x)] = f_{xx} + 2ff_{xy} + f_x f_y + f(f_y)^2 + f^2 f_{yy} \tag{8.10}$$

and so on. The equations (8.8), (8.9), (8.10) and subsequent equations can be substituted into (8.7) to give a series expansion, in h, to evaluate the solution *one* step, h, from a given point x.

It is immediately obvious that this technique has many disadvantages. First, we must be able to form the derivatives appearing in (8.7) and we can see that, even for a relatively simple function $f(x,y)$, this is a non-trivial problem. If a symbolic manipulation language is available on the computer, then it may be possible to form these derivatives, but this does not overcome the second problem: some of the partial derivatives that we have formed analytically may *not* exist at the point at which they are to be evaluated! One further problem is that we have assumed that it is reasonable to form these partial derivatives but Theorem 8.1 states that only the *first* derivative of the solution need exist.

For these reasons we shall pay no further attention to methods applying Taylor's series directly but merely note that numerical methods, similar to that described above, do exist and some are described by Lambert. However, it is very important to realise that the numerical methods that we shall describe are based on this expansion in the sense that they attempt to match terms in (8.7).

8.2 Linear multistep methods

We are trying to solve the initial-value problem (8.1)

$$y' = f(x,y)$$

$$y(a) = \eta$$

for x in the range $a \leq x \leq b$ with a and b being finite. We assume that $f(x,y)$ satisfies the conditions of Theorem 8.1 and so the solution is unique. Rather than try to obtain the solution at *every* point in $[a,b]$ we shall concentrate on determining the solution at the *set* of points $x_i = a + ih$ $(i=1,2,\ldots,m)$ where $h=(b-a)/m$. We have to be careful about notation from now on because we shall denote the *approximation* to the value of $y(x_i)$ by y_i and so y_i is not necessarily equal to $y(x_i)$, the analytic solution. Thus $f_i = f(x_i, y_i)$ is an *approximation* to $f(x_i, y(x_i))$.

A *linear multistep* method, of stepnumber k, takes the form

$$\sum_{j=0}^{k} \alpha_j y_{n+j} = h \sum_{j=0}^{k} \beta_j f_{n+j} \tag{8.11}$$

giving a *linear* relationship between the y_{n+j} and the f_{n+j}; all the α_j and β_j are constants. We shall assume that $\alpha_k \neq 0$ and that α_0 and β_0 are not both zero. We can multiply both sides of (8.11) by a constant without modifying the relationship and so, to remove this arbitrariness, we assume that $\alpha_k = 1$.

Two examples of linear multistep methods are Simpson's rule (2-step)

$$y_2 - y_0 = \frac{h}{3}(f_0 + 4f_1 + f_2) \tag{8.12}$$

and Milne's rule (4-step)

$$y_4 - y_0 = \frac{4h}{3}(2f_1 - f_2 + 2f_3) \tag{8.13}$$

In Section 8.2.1 we shall describe how these methods are derived; as we can see there are obvious similarities to the integration formulae of Section 7.2.1 which bear the same names. In Milne's rule (8.13) we can see that $\beta_4=0$ and so, given the values of y_0, y_1, y_2 and y_3 (and hence f_0, f_1, f_2 and f_3) we can calculate a value for y_4 directly. Formulae of this type are called *explicit*. In contrast, in Simpson's rule $\beta_2 \neq 0$ and so we cannot calculate y_2 directly because $f_2=f(x_2,y_2)$ is a function of the unknown value of y_2; this type of formula is said to be *implicit*. We can solve (8.12) by setting up the iteration

$$y_2^{[n+1]} = y_0 + \frac{h}{3}(f_0 + 4f_1 + f_2^{[n]})$$

and choosing a suitable value for $y_2^{[0]}$.

Convergence is guaranteed if

$$h<3/L$$

where L is the Lipschitz constant for f.

Proof

Using the theory developed in Section 3.1.4 this iteration will converge if

$$\frac{h}{3}\left|\frac{\partial f_2}{\partial y}\right| < 1$$

that is, if

$$h < \frac{3}{|\partial f_2/\partial y|}$$

Now if L is a Lipschitz constant such that

$$\left|\frac{\partial f_2}{\partial y}\right| < L$$

then

$$\frac{1}{L} < \left| \frac{1}{\partial f_2/\partial y} \right|$$

and so convergence is ensured if $h<3/L$.

Thus we have an indication from the very outset that the choice of the step length, h, can critically affect the behaviour of the numerical method.

8.2.1 Derivation of numerical methods

Through Numerical Integration
If we integrate the first derivative of y with respect to x between the limits x_0 and x_k, then we obtain

$$\int_{x_0}^{x_k} y'(x)\,\mathrm{d}x = y(x_k) - y(x_0)$$

Now, from the differential equation (8.1) we are given that

$$y' = f(x,y)$$

and so we have the result that

$$y(x_k) - y(x_0) = \int_{x_0}^{x_k} f(x,y)\,\mathrm{d}x$$

If we now vary x_k, the upper limit of the integral, and replace the analytic integration by a numerical integration, then we have a formula linking discrete values of y and $f(x,y)$. For example, if we choose $k=2$ and replace the analytic integration using Simpson's rule, then we obtain

$$y(x_2) - y(x_0) \simeq \frac{h}{3}[f(x_0,y(x_0)) + 4f(x_1,y(x_1)) + f(x_2,y(x_2))]$$

This is an approximate relationship between *exact* values of y and $f(x,y)$. Alternatively, replacing $y(x_i)$ by y_i, it can be thought of as the exact relationship

$$y_2 - y_0 = \frac{h}{3}[f_0 + 4f_1 + f_2]$$

between *approximate* values of y and f. This derivation technique limits our consideration to methods in which $\alpha_k=1$, $\alpha_j=-1$ and $\alpha_i=0$ $(i=1,2,\ldots,j-1,j+1,\ldots,k,\ j\neq k)$. The technique which follows is much more general.

Method of Undetermined Coefficients

This method is based on the use of Taylor's series

$$y(x_0+h) = y(x_0) + hy'(x_0) + \frac{h^2}{2!}\,y''(x_0) + \frac{h^3}{3!}\,y'''(x_0) + \cdots$$

One of the simplest methods is derived by truncating this series after the second term and replacing y' using the differential equation to give

$$y(x_0 + h) \simeq y(x_0) + hf(x_0,y(x_0))$$

Reinterpreting this as an exact relationship between approximate values leads to the formula

$$y_1 = y_0 + hf_0 \tag{8.14}$$

which is known as *Euler's* method. The error associated with this method is

$$\frac{h^2}{2!}\,y''(x_0) + \frac{h^3}{3!}\,y'''(x_0) + \cdots \tag{8.15}$$

which is that part of the Taylor's series removed by the truncation; for this reason it is known as the *local discretisation* or *local truncation* error. It is extremely important to realise that this is a *local* error which is specific to the step taken between x_0 and x_0+h, and is dependent on y_0 and f_0 being *exact*, not computed values. This last assumption is known as the *localising* assumption. The accumulation of all the local and rounding errors incurred at each step constitutes a *global* error. These two errors should not be confused. From (8.15) we can see that the local error is $O(h^2)$ and is identically zero if the solution is a polynomial of degree at most one. {An expression is $O(h^k)$ if it behaves like Ch^k as $h\rightarrow 0$, where C is a constant, independent of h.}

The method of undetermined coefficients consists of choosing k, the number of steps that are to be used in a formula, and then choosing the coefficients α_j $(j=0,1,\ldots,k-1)$ and β_j $(j=0,1,\ldots,k)$ in some rational fashion, usually to make the local truncation error as small as possible. We illustrate the use of this method by deriving the most accurate linear one-step method

$$y_1 + \alpha_0 y_0 = h[\beta_0 f_0 + \beta_1 f_1]$$

We write down the equivalent approximate relationship

$$y(x_0 + h) + \alpha_0 y(x_0) \simeq h[\beta_0 y'(x_0) + \beta_1 y'(x_1)] \tag{8.16}$$

Now, replacing $y(x_0+h)$ by

$$y(x_0 + h) = y(x_0) + hy'(x_0) + \frac{h^2}{2!} y''(x_0) + \cdots$$

and $y'(x_0+h)$ by

$$y'(x_0+h) = y'(x_0) + hy''(x_0) + \frac{h^2}{2!} y'''(x_0) + \cdots$$

and collecting together all the right-hand terms gives

$$C_0 y(x_0) + C_1 hy'(x_0) + C_2 h^2 y''(x_0) + C_3 h^3 y'''(x_0) + \cdots \simeq 0$$

in which

$$C_0 = 1 + \alpha_0 \qquad C_1 = 1 - \beta_0 - \beta_1$$

$$C_2 = \tfrac{1}{2} - \beta_1 \qquad C_3 = \tfrac{1}{6} - \tfrac{1}{2}\beta_1$$

Thus, in order to make (8.16) as accurate as possible we choose $\alpha_0 = -1$, $\beta_0 = \frac{1}{2}$ and $\beta_1 = \frac{1}{2}$ so that $C_0 = C_1 = C_2 = 0$. Now C_3 becomes $-\frac{1}{12}$. The one-step method is now

$$y_1 - y_0 = \frac{h}{2}(f_0 + f_1) \tag{8.17}$$

and this is the *trapezoidal* rule (which could have been obtained using the method of Section 8.2.1, p.255) and has local truncation error

$$-\tfrac{1}{12}h^3 y'''(x_0) + \cdots$$

We can use this derivation process to determine the local truncation error of a *given* method, replacing y_j by $y(x_0+jh)$ and expanding using Taylor's series.

This approach to the derivation of methods is not without problems. For example, if the series expansions were taken about a different point would different formulae result? Furthermore, we are again assuming the existence of higher derivatives of y; is this valid? These questions will be discussed in Section 8.2.3.

8.2.2 Convergence

We have to be very precise about what we mean by convergence. In the context of ordinary differential equations we want our approximation to the solution *at a particular point* to tend to the exact solution at that point as some parameter, in this case the step length, tends to zero. Now, each of the discrete points in the interval (a,b) is represented in the form

$$x = a + rh \qquad \text{for some } r>0$$

and so as h tends to zero so r must tend to infinity in order that the point x remain fixed. For example, the mid-point of the interval $(0,1)$ is $x_1=a+1h$ when $h=0.5$ but is $x_2=a+2h$ when $h=0.25$, and so on. Bearing this in mind we give the following definition of convergence from Lambert (1973, p.22).

Definition 8.2

The linear multistep method (8.11) is said to be *convergent* if, for all initial-value problems (8.1) satisfying Theorem 8.1, we have that

$$\lim_{\substack{h\to 0\\ ih=x-a}} y_i=y(x_i)$$

holds for all x in $[a,b]$, and for all solutions $\{y_i\}$ of (8.11) satisfying the initial condition

$$y_\mu = \eta_\mu(h)$$

for which

$$\lim_{h\to 0} \eta_\mu(h) = \eta \qquad \text{for } \mu=0,1\,.\,.\,.,k-1$$

One further point to note in this definition is that part concerned with the initial condition. The multistep method (8.11) involves $k+1$ y-values, k of which must be available in order that y_{n+k} can be calculated; i.e. k starting values must be available. Thus, as h tends to zero these starting values will come from points closer and closer to the initial point a and so, for convergence to take place, these additional values must tend to the starting value, η. A great deal of freedom remains; the starting values do not have to be exact solutions of (8.1), in fact we do not even need $y(a)=\eta$, although we would normally make this choice.

8.2.3 Order

In this section we shall answer the questions posed at the end of Section 8.2.1 (p.256). To do this we define the linear operator $\mathcal{L}$ by

$$\mathcal{L}[y(x);h] = \sum_{j=0}^{k} [\alpha_j y(x+jh) - h\beta_j y'(x+jh)] \tag{8.18}$$

where, now, $y(x)$ is an arbitrary function, continuously differentiable on $[a,b]$. Because $y(x)$ is arbitrary we can assume that it has as many continous derivatives as we require and so we can analyse the formula (8.11), appearing in (8.18), without worrying about the solution to a *particular* differential equation.

Using Taylor's series, as described in Section 8.2.1 (p.256), we can rewrite (8.18) in the form

$$\mathcal{L}[y(x);h] = C_0 y(x) + C_1 h y'(x) + \cdots + C_i h^i y^{(i)}(x) + \cdots \tag{8.19}$$

where the C_i are constants. We can now give a definition of the order of any particular method.

Definition 8.3

The operator (8.18) and the associated linear multistep method (8.11) are said to be of *order p* if, in (8.19), $C_0 = C_1 = \cdots = C_p = 0$ and $C_{p+1} \neq 0$.

It is a relatively simple task to show that the constants C_i in (8.19) have the form

$$\begin{aligned} C_0 &= \alpha_0 + \alpha_1 + \cdots + \alpha_k \\ C_1 &= \alpha_1 + 2\alpha_2 + \cdots + k\alpha_k - (\beta_0 + \beta_1 + \cdots + \beta_k) \\ C_i &= \frac{1}{i!}(\alpha_1 + 2^i \alpha_2 + \cdots + k^i \alpha_k) \\ &\quad - \frac{1}{(i-1)!}(\beta_1 + 2^{i-1}\beta_2 + \cdots + k^{i-1}\beta_k) \qquad \text{for } i=2,3,\ldots \end{aligned} \tag{8.20}$$

We can use these formulae to derive linear multistep methods of a given step number and maximal order. For example, compare these equations with those used in Section 8.2.1 (p.256) to derive the trapezoidal rule.

We have yet to answer the question posed in Section 8.2.1: 'Does the order of the linear multistep method change if the Taylor's series expansions are based on a different point?' The following theorem calms our disquiet.

Theorem 8.2

The order of the linear multistep method (8.11) does not change if the Taylor's series is expanded about a different point.

Proof

We suppose that the terms in (8.18) are expanded using Taylor's series, but this time about the point $x+rh$ $\{r\neq 0\}$. As a result (8.19) takes the form

$$\mathcal{L}[y(x);h] = D_0 y(x+rh) + D_1 h y'(x+rh) + \cdots + D_i h^i y^{(i)}(x+rh) + \cdots \tag{8.21}$$

Now each of the terms on the right-hand side of this equation can be expanded about the point x, again using Taylor's series, and the results equated to (8.19). This gives the conclusion that $C_0=C_1=\cdots=C_p=0$ if and only if $D_0=D_1=\cdots=D_p=0$ and so $\mathcal{L}$ has order p if the first $(p+1)$ terms in (8.21) or (8.19) are zero, independently of the choice of r. Further, if $C_0=C_1=\cdots=C_p=0$, then $D_{p+1}=C_{p+1}$ but $D_{p+2}, D_{p+3}, \ldots$ are still functions of r. For this reason only the first non-vanishing term is of practical significance because the rest are dependent on the choice of r. This non-vanishing coefficient C_{p+1} is called the *error constant*. ■

8.2.4 Consistency and zero-stability

Definition 8.4

The linear multistep method (8.11) is said to be *consistent* if its order, p, is at least one.

If we consider equations (8.20) we see that (8.11) is consistent if $C_0=C_1=0$ and so

$$\alpha_0 + \alpha_1 + \cdots + \alpha_k = 0$$

and

$$\alpha_1 + 2\alpha_2 + \cdots + k\alpha_k = \beta_0 + \beta_1 + \cdots + \beta_k$$

If the formula (8.11) satisfies these two equations then it will exactly integrate a differential equation whose solution is a linear polynomial. Ralston and Rabinowitz (1978, p.177) prove the following theorem:

Theorem 8.3

Consistency is necessary for convergence.

However, Lambert (1973, p.32) shows by an example that consistency is *not sufficient* for convergence.

In order to define the extra condition necessary for convergence we define the *first characteristic polynomial* of (8.11) to be

$$\rho(\xi) = \sum_{j=0}^{k} \alpha_j \xi^j$$

and the *second* to be

$$\sigma(\xi) = \sum_{j=0}^{k} \beta_j \xi^j$$

The solution of equation (8.11) comprises two parts, one defined by the left-hand side of the equation and the other by the right-hand side. Lambert (1973, p.32) shows that the part associated with the left-hand side has the form

$$y_m = \gamma_1 \xi_1^m + \gamma_2 \xi_2^m + \cdots + \gamma_k \xi_k^m$$

where $\xi_1, \xi_2, \ldots, \xi_k$ are the roots of the first characteristic polynomial, $\rho(\xi)$. If the formula (8.11) is consistent, then it is immediately obvious that

$$\rho(1) = 0$$

and

$$\rho'(1) = \sigma(1)$$

From the first equation it follows that $\rho(\xi)$ has a root at $\xi=1$; this root is known as the *principal* root and is denoted by ξ_1. The remaining roots are known as *spurious* roots and, obviously, exist only if $k>1$. In this case we have replaced the first-order equation (8.1a) by a *difference* equation (8.11) of order $k>1$ and the roots $\xi_2, \xi_3, \ldots, \xi_k$ do not represent, in any way, the behaviour of the solution of the differential equation. It is therefore vital that these components of the solution are damped out as h becomes smaller. This will happen if none of these roots has a modulus greater than 1. Further, it is essential that all roots of modulus 1 are simple (*not* multiple) roots. These conditions are encompassed in the following definition.

Definition 8.5

The linear multistep method (8.11) is *zero-stable* if no root of the first characteristic polynomial $\rho(\xi)$ has modulus greater than one, and if every root of modulus one is simple (i.e. not a repeated root).

The concepts of consistency, zero-stability and convergence are drawn together in the following theorem, proved in Henrici (1962).

Theorem 8.4

The necessary and sufficient conditions for a linear multistep method to be convergent are that it be consistent and zero-stable.

A proof of the following theorem concerning the maximum possible order that a zero-stable linear k-step method can attain also appears in Henrici.

Theorem 8.5

No zero-stable linear k-step method can have order exceeding $k+1$ when k is odd, or exceeding $k+2$ when k is even.

This theorem confirms the result of our manipulation in Section 8.2.1 (p.256) that the trapezoidal rule, a one-step method of order two, has maximal order.

8.2.5 Absolute and relative stability

Zero-stability, discussed in the previous section, is relevant only in situations where $h \to 0$. In reality, we are interested in the behaviour of the methods for a fixed value of h as more and more steps are taken. Now, the important consideration is what is happening to the total (or *global*) error, the accumulation of truncation and rounding errors incurred at each step, as the calculation progresses. We shall be looking for criteria applicable to the methods to determine whether or not this global error is controlled. The discussion which follows is strongly based on that given in Lambert (1973, p.64, *et seq.*).

We shall illustrate the criteria with reference to the test differential equation

$$y' = \lambda y \qquad \lambda \text{ constant} \tag{8.22}$$

whose solution, given the initial condition $y(a)=\eta$, is

$$y(x) = \eta e^{\lambda(x-a)}$$

If $\lambda>0$ then the solution increases and if $\lambda<0$ the solution decreases. We justify using this particular differential equation by noting that, *locally*, any first-order ordinary differential equation

$$y' = f(x,y)$$

can be linearised to give

$$y' = \alpha y + \beta$$

where α is an approximation to $\partial f/\partial y$. This equation behaves, essentially, like (8.22).

We now consider the general k-step method, (8.11),

$$\sum_{j=0}^{k} \alpha_j y_{n+j} = h \sum_{j=0}^{k} \beta_j f_{n+j}$$

which we assume to be consistent and zero-stable. The *theoretical* solution of the differential equation (8.22) satisfies

$$\sum_{j=0}^{k} \alpha_j y(x_{n+j}) = h\lambda \sum_{j=0}^{k} \beta_j y(x_{n+j}) + T_{n+k} \tag{8.23}$$

where T_{n+k} is the local truncation error. If we denote by $\tilde{y}_n$ the solution of (8.11) when a rounding error R_{n+k} is committed at the nth application of the method, then

$$\sum_{j=0}^{k} \alpha_j \tilde{y}_{n+j} = h\lambda \sum_{j=0}^{k} \beta_i \tilde{y}_{n+i} + R_{n+k} \tag{8.24}$$

Subtracting (8.24) from (8.23) gives the equation

$$\sum_{j=0}^{k} \alpha_j [y(x_{n+j}) - \tilde{y}_{n+j}] = h\lambda \sum_{j=0}^{k} \beta_j [y(x_{n+j}) - \tilde{y}_{n+j}] + \phi_{n+k}$$

where $\phi_{n+k}=T_{n+k}-R_{n+k}$. Making the assumption that $\phi_{n+k}=\phi=$constant and defining $\tilde{\epsilon}_{n+j}=y(x_{n+j}) - \tilde{y}_{n+j}$ this can be rewritten as

$$\sum_{j=0}^{k} (\alpha_j - h\lambda\beta_j)\tilde{\epsilon}_{n+j} = \phi$$

The general solution of this equation

$$\tilde{\epsilon}_n = \sum_{j=0}^{k} d_j r_j^n - \phi / \left(h\lambda \sum_{j=0}^{k} \beta_j \right) \tag{8.25}$$

is given in Lambert. The d_j (j=0,1,. . .,k) are arbitrary constants and the r_j (j=0,1,. . .,k) are the roots of the *stability* equation

$$\sum_{j=0}^{k} (\alpha_j - h\lambda\beta_j) r^j = 0 \tag{8.26}$$

Although it is not necessary for the results which follow to be valid, these roots are assumed to be distinct. Using the notation of Section 8.2.4, equation (8.26) can be rewritten in the form

$$\pi(r,\bar{h}) = \rho(r) - \bar{h}\sigma(r) = 0 \tag{8.27}$$

where we have defined $\bar{h}=h\lambda$; thus $\pi(r,\bar{h})$ is the *stability polynomial* of the linear multistep method (8.11).

We are not interested in determining the exact value, or even the magnitude, of $\tilde{\epsilon}_n$ but merely in attempting to discover whether or not $\tilde{\epsilon}_n$ grows as n increases. The following definition can be found in Lambert (1973, p.66).

Definition 8.6

The linear multistep method (8.11) is said to be *absolutely stable* for a given $\bar{h}$ if, for that $\bar{h}$, all the roots of (8.27) satisfy

$$| r_i | <1 \qquad r = 0,1, \ldots ,k$$

and to be *absolutely unstable* for that $\bar{h}$ otherwise. An interval (α,β) is said to be an *interval of absolute stability* if the method is absolutely stable for all $\bar{h}$ in (α,β). If the method is absolutely unstable for all $\bar{h}$, then it is said to have *no interval of absolute stability*.

The interval of absolute stability is dependent only on the coefficients of the method, but the choice of an h for which $\tilde{\epsilon}_n$ will not increase is dependent on λ and so on the particular differential equation. We note, in passing, that this means that different values of h may be necessary in different regions; the differential equation (8.22) approximates the behaviour of a general differential equation only locally and the value of λ, used in the linearisation, probably needs to change from region to region. If λ becomes large then, to control the magnitude of $\bar{h}=h\lambda$, the maximum value of h may have to be reduced.

When $\bar{h}=0$ then $\pi(r,0)=\rho(r)$ and the roots of the stability polynomial are those of the first characteristic polynomial of the method. By assuming that the method is consistent and zero-stable we have ensured that $\rho(r)$ has a simple root at $r=1$ (see Section 8.2.4). Let r_1 be the root of $\pi(r,\bar{h})$ which tends to $+1$ as $\bar{h}\rightarrow 0$, then Lambert (1973, p.66) proves the result

Lemma 8.1

$$r_1 = e^{\bar{h}} + O(\bar{h}^{p+1}) \qquad \text{as } \bar{h}\rightarrow 0 \tag{8.28}$$

where p is the order of the linear multistep method.

If we consider (8.28) we can see immediately that, for small positive $\bar{h}$, r_1 is greater than 1 and so we conclude that *every consistent zero-stable linear* *multistep method is absolutely unstable for small positive $\bar{h}$.* Thus the interval of absolute stability must always be of the form $(\alpha,0)$. There do exist methods for which α is zero.

The fact that the interval of absolute stability is always negative need not be a handicap. If $\bar{h}>0$, then this implies that $\lambda>0$ and so the solution of (8.22) is the increasing exponential $e^{\lambda(x-a)}$. From (8.25), the solution of the error equation contains a term r_1^n and from (8.28) this is

$$r_1^n = (e^{\bar{h}})^n + O(\bar{h}^{p+1})$$

Now,

$$(e^{\bar{h}})^n = e^{\lambda(x_n-a)}$$

and so it follows that, if the solution of the error equation is dominated by the term in r_1^n, the error grows at a similar rate to that of the solution of the differential equation and this situation is usually considered to be acceptable. Therefore, there is another definition of stability (Lambert, 1973, p.68).

Definition 8.7

The linear multistep method (8.11) is said to be *relatively stable* for a given $\bar{h}$ if, for that $\bar{h}$, the roots r_i of (8.27) satisfy

$$|r_i| < |r_1| \qquad \text{for } i = 2,3\ldots,k$$

and to be *relatively unstable* otherwise. An interval (α,β) is said to be an *interval of relative stability* if the method is relatively stable for all $\bar{h}$ in (α,β).

A method which is relatively stable for small negative $\bar{h}$ is also absolutely stable for that $\bar{h}$ because, from (8.28), $|r_1| <1$. For this reason relative stability may be thought to be the more useful concept. However, the analysis of the relative stability properties of methods is more difficult than that of the absolute stability properties. Also, there are situations, concerning systems of ordinary differential equations, where absolute stability is more relevant.

We are left with the problem of determining the intervals of absolute and/or relative stability. We shall discuss only one method, the *root-locus* method, and that briefly; the interested reader will find other techniques discussed in Lambert. The root-locus method proceeds by determining the roots of equation (8.27) for a range of values of $\bar{h}$; the procedure *Bairstow* of Fig. 3.16 is appropriate for this task. A graph is drawn of $|r_i|$ $\{i=1,2,\ldots,k\}$ against $\bar{h}$ and, from it, an attempt is made to deduce intervals of stability.

We illustrate the importance of the ideas of absolute and relative stability by considering Simpson's method (8.12). The stability equation is

$$(1-\tfrac{1}{3}\bar{h})r^2 - \tfrac{4}{3}\bar{h}r - (1+\tfrac{1}{3}\bar{h}) = 0$$

and Lambert performs the analysis to show that, to $O(\bar{h}^2)$, the two roots are given by

$$r_1 = 1 + \bar{h}$$

$$r_2 = -1 + \tfrac{1}{3}\bar{h}$$

We can see immediately that there is no interval of absolute stability; if $\bar{h}>0$ then $r_1>1$ and if $\bar{h}<0$ then $r_2<-1$. However, there is an interval of relative stability of the form $(0,\beta)$ and it is possible to show that this interval is, in fact, $(0,\infty)$. We conclude, therefore, that if $\bar{h}<0$ $(\lambda<0)$ then Simpson's rule is not appropriate, whereas if $\bar{h}>0$ $(\lambda>0)$ the errors will not grow relative to the solution. The values in Table 8.1 were produced by applying Simpson's method, with $h=0.1$, to the initial-value problem

$$y' = \lambda y$$

$$y(0) = 1$$

for the case $\lambda=5$. The error displayed in the table is $\hat{y}_n - y(x_n)$, the difference between the numerical solution and the analytic solution.

Table 8.1

x	$\hat{y}_n$	$y(x_n)$	*Error*
0.0	1.0000	1.0000	0.0000
0.2	2.0407	2.7183	1.3404(−1)
0.4	7.6580	7.3891	2.6890(−1)
0.6	2.0760(+1)	2.0086(+1)	6.7435(−1)
0.8	5.6404(+1)	5.4598(+1)	1.8063
1.0	1.5334(+2)	1.4841(+2)	4.9286
1.5	1.8695(+3)	1.8080(+3)	6.1470(+1)
2.0	2.2795(+4)	2.2027(+4)	7.6815(+2)
2.5	2.7793(+5)	2.6834(+5)	9.5924(+3)

The values in Table 8.1 are consistent with our claim that Simpson's method is relatively stable for small positive $\bar{h}$; although the error is becoming larger it remains noticeably smaller in magnitude than the solution itself.

The values in Table 8.2 result from choosing $\lambda = -5$ and are displayed for x-values which differ from those in Table 8.1 to illustrate the rapid deterioration for $\bar{h}<0$. At $x=0.2$ the computed value differs little from the analytic solution but the error subsequently oscillates in sign and becomes larger. At $x=0.5$ the error is already of the same order of magnitude as the solution and by $x=1$ the analytic solution is completely swamped.

Table 8.2

x	$\hat{y}_n$	$y(x_n)$	*Error*
0.0	1.0000	1.0000	0.0000
0.2	4.2391(−1)	3.6788(−1)	5.6027(−2)
0.3	1.9100(−1)	2.2313(−1)	−3.2126(−2)
0.4	1.9365(−1)	1.3534(−1)	5.8310(−2)
0.5	2.5777(−2)	8.2085(−2)	−5.6308(−2)
1.0	1.4424(−1)	6.7379(−3)	1.3751(−1)
1.5	−3.1038(−1)	5.5308(−4)	−3.1094(−1)
2.0	7.0502(−1)	4.5400(−5)	7.0498(−1)
2.5	−1.5982	3.7268(−6)	−1.5982

8.3 Predictor–corrector methods

At the beginning of Section 8.2 we pointed out that the solution of an implicit equation, in that case Simpson's method, could be obtained using an iteration provided that the step length h satisfied a certain condition. However, we did not explain how this iteration was to be implemented. How, for example, are we to obtain an initial approximation to the unknown value and, moreover, where are the previous y-values to come from? We illustrate the answers to these questions by considering a particular formula, Simpson's formula,

$$y_2 - y_0 = \frac{h}{3}(f_0 + 4f_1 + f_2) \tag{8.12}$$

We shall start with the second question; where do the starting values come from? In order to calculate y_2, in (8.12), we need values for y_0, f_0 and f_1. If we assume that we are at the very beginning of the calculation then $x_0 = a$ and so $y_0 = \eta$ and $f_0 = f(a, \eta)$. We must now find a value for f_1. We can use the Euler method, described earlier, which is a one-step explicit method, but unfortunately this method is inaccurate and so is not really appropriate. If h is sufficiently small, then we can use Taylor's series methods, but we have already discussed their disadvantages. What we require is an *accurate* one-step method and such methods (Runge–Kutta methods) are the topic of Section 8.4. So, let us now assume that we have the necessary starting values and what remains is for us to determine an initial approximation to y_2 to substitute into the iteration

$$y_2^{[n+1]} = y_0 + \frac{h}{3}(f_0 + 4f_1 + f_2^{[n]}) \tag{8.29}$$

The convergence of this iteration will be faster if we provide a good initial estimate, $y_2^{[0]}$, and the standard method of obtaining such an initial approximation is to use an explicit method to *predict* $y_2^{[0]}$. We then use (8.29) to *correct* it. Thus a *predictor–corrector* method comprises two multistep methods, an explicit one to predict the new value and an implicit one to correct it. For example, Milne's rule (8.13) is often used as the predictor for Simpson's method giving the predictor–corrector pair

$$y_4^{[0]} - y_0 = \frac{4h}{3}(2f_1 - f_2 + 2f_3)$$

and

$$y_4^{[n+1]} - y_2 = \frac{h}{3}(f_2 + 4f_3 + f_4^{[n]})$$

We note that this pair of methods requires four starting values y_0, y_1, y_2 and y_3.

We now introduce some notation to help describe the way that predictor–corrector methods are used in practice. The letter **P** denotes an application of the predictor, **C** the application of the corrector and **E** the evaluation of the function $f(x,y)$. If the iteration (8.29) proceeds until two successive iterates agree to within some specified tolerance, then it is said to have *iterated to convergence*. We denote the general situation by

$$\mathbf{P(EC)}^m\mathbf{E} \tag{8.30}$$

that is, a prediction, an evaluation of f and a correction repeated m times (possibly infinitely) and a final evaluation of the function f (this final function evaluation enhances the stability of the method). The expression (8.30) represents *one* step of the predictor–corrector method. It is extremely rare for predictor–corrector methods to be used in this manner; their most common implementations are in the modes **PEC** or **PECE**. The motivation for these implementations is to keep the number of function evaluations to a minimum but, obviously, this will have some effect on the solutions produced.

If we iterate to convergence as in (8.30) then, provided that the predictor is convergent, all the properties of the predictor–corrector pair are those of the corrector alone. However, if iteration to convergence is not used, then the properties of the method depend on *both* the predictor and the corrector and these have to be matched very carefully. The choice depends critically on the orders and stability properties of the methods. The analysis governing this choice is beyond the scope of this book; the interested reader is referred to Lambert (1973, Section 3.9, *et seq*).

The class of *Adams–Bashforth–Moulton* predictor–corrector methods is particularly popular. Adams methods all have the property that

$$\rho(\xi) = \xi^k - \xi^{k-1}$$

that is

$$\alpha_k = 1 \qquad \text{and} \qquad \alpha_{k-1} = -1$$

with $\alpha_j=0$ $\{j=0,1, \ldots, k-2\}$. Thus the principle root of the first characteristic polynomial is $\xi_1=+1$ and the others are all zero and so the Adams methods generally have substantial intervals of absolute stability. The Adams–Bashforth methods are the explicit members of the class and the Adams–Moulton methods are the implicit ones. The following pair of methods from the class from a popular predictor–corrector pair

$$\mathbf{P}: \quad y_4 - y_3 = \frac{h}{24}(55f_3 - 59f_2 + 37f_1 - 9f_0)$$

$$\mathbf{C}: \quad y_4 - y_3 = \frac{h}{24}(9f_4 + 19f_3 - 5f_2 + f_1)$$

and constitute a fourth-order method. The procedure of Fig. 8.2 implements this pair of methods in the **PECE** mode to perform one step of the integration of the ordinary differential equation

$$y' = f(x,y)$$

The procedure assumes that the necessary starting values have been previously calculated.

```
procedure ABM (function f (x, y : real) : real;
                    h, x4, y3, f0, f1, f2, f3 : real;
                    var y4, f4 : real);
  var
    predy4, predf4 : real;
begin
  predy4 := y3 + h * (55 * f3 - 59 * f2 + 37 * f1 - 9 * f0)/24;
  predf4 := f(x4, predy4);
  y4 := y3 + h * (9 * predf4 + 19 * f3 - 5 * f2 + f1)/24;
  f4 := f(x4, y4)
end {Adams-Bashforth-Moulton};
```

Figure 8.2

Table 8.3 shows the values computed by this procedure when used to solve the initial-value problem

$$y' = e^{2x} + y, \qquad y(0) = 1$$

at the equidistant points $x = 0.4, 0.5, \ldots, 1.0$, given the starting values shown. All the values in the last two columns are computed from the exact solution

$$y = e^{2x}$$

Table 8.3

x	P	E	C	E	y	f
0.0					1.000000	2.000000
0.1					1.221403	2.442806
0.2					1.491825	2.983649
0.3					1.822119	3.644238
0.4	2.225375	4.450916	2.225549	4.451090	2.225541	4.451082
0.5	2.718090	5.436372	2.718301	5.436583	2.718282	5.436564
0.6	3.319892	6.640009	3.320151	6.640268	3.320117	6.640234
0.7	4.054937	8.110137	4.055253	8.110453	4.055200	8.110400
0.8	4.952724	9.905756	4.953110	9.906143	4.953032	9.906065
0.9	6.049284	12.098932	6.049756	12.099404	6.049647	12.099295
1.0	7.388628	14.777684	7.389205	14.778261	7.389056	14.778112

8.3.1 Error control

We have already noted that the choice of step length, h, must be made carefully if the iteration (8.29) is to converge. This choice is equally critical if we use predictor–corrector methods in the mode **PECE**; in this case the choice of h more strongly affects the accuracy and stability of the method. Again, discussion of the stability aspects of this choice are beyond the scope of this book, but we can outline the process whereby the step length is chosen to control the growth of errors (and here we are talking only about local errors). The situation is analogous to that of Section 7.2.5 in which we described the adaptive quadrature approach to the integration of a function $f(x)$. There we chose to refine (or partition) subintervals in the light of the *local* behaviour of the truncation error of the integral approximation; if the function varied rapidly, and so had large derivatives affecting the truncation error, then the interval size was reduced. We can make similar decisions, based on the local truncation error of the predictor–corrector method, for the integration of the differential equation; if the solution is varying rapidly then we will reduce the step length, if it is not, then instead we can increase the step length. The technique that we shall use to obtain an estimate of this local truncation error is known as *Milne's device*.

We assume that both the predictor and the corrector have the same order, p, and that the principal local truncation error of the corrector is given by

$$C_{p+1}h^{p+1}y^{(p+1)}(x_0) = y(x_k) - y_k^{[m]} + O(h^{p+2}) \tag{8.31}$$

and that of the predictor by

$$C^*_{p+1}h^{p+1}y^{(p+1)}(x_0) = y(x_k) - y_k^{[0]} + O(h^{p+2}) \tag{8.32}$$

The superscript $[m]$ denotes the mth iterate of the predictor–corrector pair used in the mode **P(EC)**^{m}**E** and [0] identifies the initial estimate provided by the predictor alone. Subtracting (8.31) from (8.32) we have

$$(C^*_{p+1} - C_{p+1})h^{p+1}y^{(p+1)}(x_0) = y_k^{[m]} - y_k^{[0]} + O(h^{p+2})$$

and so, rearranging this equation and using (8.31), it follows that the principal local truncation error of the method, that of the corrector, is estimated by

$$C_{p+1}h^{p+1}y^{(p+1)}(x_0) = \frac{C_{p+1}}{C^*_{p+1} - C_{p+1}}(y_k^{[m]} - y_k^{[0]}) \tag{8.33}$$

It is emphasized again that this result is dependent upon the localising assumption that all previous values, used to calculate $y^{[m]}$ and $y^{[0]}$, are exact. Of course, this is not the case in practice, but we hope that (8.33) will still provide a reasonable estimate of the principal part of the local truncation error.

We illustrate the use of (8.33) by considering the Milne–Simpson method described earlier. For this pair the order is 5 and for the corrector, Simpson's method, $C_5 = -1/90$ and, for the predictor, Milne's method, $C_5^* = 14/45$. Substituting these values into (8.33) gives the result

$$C_5h^5y^{\text{v}}(x_0) = \frac{-\frac{1}{90}}{\frac{14}{45} + \frac{1}{90}}(y_4^{[m]} - y_4^{[0]}) = -\tfrac{1}{29}(y_4^{[m]} - y_4^{[0]})$$

If the modulus of the right-hand side of this equation is less than ϵ/h^5, where ϵ is some prescribed error-per-step tolerance, then we can increase the step length for the next step; if it is greater than ϵ, then we will have to reduce the step length and recalculate the solution. Examining (8.33) we see that we have produced this error estimate with almost no extra calculation; the error constants C^*_{p+1} and C_{p+1} are already known and both $y^{[m]}$ and $y^{[0]}$ have already been calculated to form the solution to the differential equation. So the only extra work necessary is the combination of these values in (8.33).

Using the starting values from Table 8.3 the Milne–Simpson method gives the following values (at $x = 0.4$)

$y^{[0]} = 2.2253919$
$f^{[0]} = 4.4509328$
$y^{[1]} = 2.2255428$
$f^{[1]} = 4.4510837$

which results in the error estimate

$$| C_5h^5y^v(x_0) | \simeq \tfrac{1}{29}(2.2255428 - 2.2253919) = 0.0000052032$$

The actual error is approximately 0.000002 and we can see that our estimate bounds this value.

In this discussion we have talked about changing the step length without considering whether this is simple to achieve. First, we consider increasing it. If we merely take an integer multiple of the existing step length, then no problem exists because the necessary starting values will already have been calculated and may have been stored, with an increase in step length in mind. For example, if we are using Simpson's method

$$y_4 - y_2 = \frac{h}{3}(f_2 + 4f_3 + f_4)$$

as corrector then, if we double the step length, the formula becomes

$$y_4 - y_0 = \frac{2h}{3}(f_0 + 4f_2 + f_4)$$

which uses values calculated with the previous step length. If we use a non-integral multiple of h, then we do have a problem because the necessary starting values will not be available. This is also the case if the step length is reduced; even if we only halve the step length we still need to calculate new values. For example, again considering Simpson's method, if we halve the step length, then the formula becomes

$$y_4 - y_3 = \frac{0.5h}{3}(f_3 + 4f_{7/2} + f_4)$$

and we need to calculate a value for $f_{7/2}$. Lambert discusses this problem and illustrates several methods for obtaining these necessary values, pointing out the difficulties occurring with each method. This topic is too advanced to discuss here and we merely note that there are difficulties in changing step length using linear multistep methods. This difficulty does not exist, in the same form, for the methods we describe in the next section.

8.4 Runge–Kutta methods

We have seen that changing the step size for linear multistep methods creates problems. These problems will not exist if only one step is taken, for then only the values at the beginning of the step need be stored; information from previous steps is not used. The simplest one-step method is Euler's method

$$y_1 = y_0 + hf_0 \tag{8.14}$$

It is an explicit method and requires no starting values other than y_0. Of course, it is not very accurate and so is of little practical use. The Taylor's series methods of Section 8.1.1 are also one-step methods and, theoretically at least, we can obtain arbitrary accuracy by including more terms in the series; however, we have already decided that there are too many problems associated with their use. *Runge–Kutta* methods are one-step methods which are designed to encompass the accuracy of the Taylor's series methods, but without evaluating the derivatives. To do this these methods sacrifice the linearity of the linear multistep methods and Runge–Kutta methods are then a special case of the general *explicit* one-step method

$$y_1 - y_0 = h\phi(x_0, y_0, h) \tag{8.34}$$

We are restricting our attention to explicit one-step methods; implicit one-step methods do exist (the trapezoidal method is one) and have properties which are useful for solving particular classes of ordinary differential equations, but the analysis of these methods is beyond the scope of this book and so we shall not discuss them further.

The definition (8.34) of the general method contains no mention of the function $f(x,y)$ from the differential equation and so our definition of the order of the method cannot be independent of the differential equation as was the case with the linear multistep methods of Section 8.2.3.

Definition 8.8

The method (8.34) is said to have *order p* if p is the largest positive integer for which

$$y(x+h) - y(x) - h\phi(x, y(x), h) \tag{8.35}$$

is $O(h^{p+1})$, where $y(x)$ is the theoretical solution of the initial-value problem.

Definition 8.9

The method (8.34) is said to be *consistent* with the initial-value problem if

$$\phi(x,y,0) \equiv f(x,y) \tag{8.36}$$

Using Taylor's series, we can rewrite (8.35) to give

$$y(x+h) - y(x) - h\phi(x,y(x),h) = hy'(x) - h\phi(x,y(x),0)$$

correct to $O(h^2)$. Now, if the method is consistent then, from (8.36),

$$\phi(x,y,0) \equiv f(x,y) \equiv y'$$

and so

$$y(x+h) - y(x) - h\phi(x,y(x), h) = O(h^2)$$

and so has order *at least* one.

Henrici (1962) proves the following theorem concerning the convergence of the method (8.34).

Theorem 8.6

(i) let $\phi(x,y,h)$ be continuous in its three arguments in the region D defined by $a \leq x \leq b$, $-\infty < y < \infty$ and $0 \leq h \leq h_0$ for some $h_0 > 0$;

(ii) let $\phi(x,y,h)$ satisfy a Lipschitz condition of the form

$$| \phi(x,y_1,h) - \phi(x,y_2,h) | \leq M | y_1 - y_2 |$$

for all points (x,y_1,h) and (x,y_2,h) in D.

Then the method (8.34) is *convergent* if and only if it is consistent.

We note an immediate difference from the convergence theorem for linear multistep methods; there is no mention of zero-stability. Difference equations for one-step methods allow only one solution, and so there is no need to worry about spurious solutions.

All the methods that we shall discuss satisfy conditions (i) and (ii) of Theorem 8.6 provided that the function $f(x,y)$ satisfies the Lipschitz condition (8.2) of Theorem 8.1.

8.4.1 Derivation of Runge–Kutta methods

The general r-stage Runge–Kutta method has the form

$$y_1 - y_0 = h\phi(x_0,y_0,h) \tag{8.37a}$$

$$\phi(x_0,y_0,h) = \sum_{i=1}^{r} \gamma_i K_i \tag{8.37b}$$

where

$$K_1 = f(x_0,y_0) \tag{8.38a}$$

$$K_i = f\left(x+h\alpha_i, y_0+h\sum_{j=1}^{i-1} \beta_{ij}K_j \right) \tag{8.38b}$$

$$\alpha_i = \sum_{j=1}^{i-1} \beta_{ij} \qquad \text{for } i=2,3, \ldots, r \tag{8.38c}$$

We can see that an r-stage Runge–Kutta method involves r evaluations of the function $f(x,y)$.

Each of the functions $K_i(x,y,h)$ $(i=1,2,\ldots,r)$ can be interpreted as an approximation to the derivative $y'(x)$ and $\phi(x,y,h)$ as a weighted average of these approximations. If we substitute $h=0$ into each equation of (8.38), then we see that each K_i $(i=1,2,\ldots,r)$ reduces to $f(x_0,y_0)$, and if we then substitute these into (8.37b) we obtain

$$\phi(x_0,y_0,0) = \sum_{i=1}^{r} \gamma_i f(x_0,y_0) = f(x_0,y_0) \sum_{i=1}^{r} \gamma_i$$

and so for the method to be consistent we require that

$$\sum_{i=1}^{r} \gamma_i = 1$$

We derive Runge–Kutta methods by insisting that the coefficients γ_i, α_i, and β_{ij} are chosen so that the formula (8.37a) agrees, as far as possible, with the Taylor's series expansion of the solution, about the point x_0. We illustrate this process by deriving a two-stage Runge–Kutta method. First we recall that, in Section 8.1.1, we wrote down the Taylor's series expansion about x_0 of $y(x_0+h)$, the solution of the differential equation $y'=f(x,y)$; substituting (8.9) and (8.10) into (8.7) we obtain

$$y_1 = y_0 + hf + \frac{h^2}{2!}(f_x+ff_y) + \frac{h^3}{3!}(f_{xx}+2ff_{xy}+f_xf_y+f(f_y)^2+f^2f_{yy}) + \cdots \tag{8.39}$$

where we assume that $f(x,y)$ and its derivatives are evaluated at the point $(x_0,y(x_0))$. It is to this series that we must match the Runge–Kutta methods.

The general two-stage Runge–Kutta method takes the form

$$y_1 = y_0 + h(\gamma_1K_1+\gamma_2K_2) \tag{8.40a}$$

$$K_1 = f(x_0,y_0) \tag{8.40b}$$

$$\begin{aligned} K_2 &= f(x_0+h\alpha_2, y_0+h\alpha_2K_1) \\ &= f(x_0+h\alpha_2, y_0+h\alpha_2 f(x_0,y_0)) \end{aligned} \tag{8.40c}$$

Equation (8.40c) is not exactly in the form (8.38b); we have taken account of the constraint (8.38c) in its construction. Before we can match (8.40) to (8.39) we need to expand K_2 using Taylor's series, this time using the two-dimensional form

$$g(x+h,y+k) = g(x,y) + (hg_x + kg_y) + \frac{1}{2!}(h^2g_{xx}+2hkg_{xy}+k^2g_{yy})$$

$$+ \text{(terms of higher order in } h \text{ and } k)$$

Applying this expansion to (8.40c) we obtain

$$K_2 = f + h\alpha_2(f_x+ff_y) + \frac{h^2\alpha_2^2}{2!}(f_{xx}+2ff_{xy}+f^2f_{yy}) + O(h^3) \tag{8.41}$$

and now, substituting (8.40b) and (8.41) into (8.40a), we have

$$y_1 = y_0 + h(\gamma_1 + \gamma_2)f + h^2\gamma_2\alpha_2(f_x+ff_y)$$

$$+ \frac{h^3\gamma_2\alpha_2^2}{2!}(f_{xx}+2ff_{xy}+f^2f_{yy}) + \cdots \tag{8.42}$$

We now compare the expansions (8.39) and (8.42) and attempt to match them term by term. The first terms, y_0, are obviously the same and the first terms involving f are the same if we choose

$$\gamma_1 + \gamma_2 = 1 \tag{8.43}$$

We notice immediately that this is the condition that *must* be satisfied if the method is to be consistent. We can also match the first derivative terms by choosing

$$\gamma_2\alpha_2 = \tfrac{1}{2} \tag{8.44}$$

and now we can see that we cannot possibly match the second derivative terms and so we stop. Thus, if we can satisfy (8.43) and (8.44) we will have constructed a second-order two-stage Runge–Kutta method (second-order because it is the term in h^3 that cannot be matched). These two equations involve the *three* unknown values γ_1, γ_2 and α_2 and, from our knowledge of linear algebra, we conclude that there exists an infinitive number of solutions; in fact, there is an infinite one-parameter family of solutions. If we choose α_2 to be the parameter and denote it by c, then (8.43) remains unchanged and (8.44) takes the form

$$\gamma_2 = \frac{1}{2c}$$

and by taking different values for c we can define different second-order two-stage Runge–Kutta methods. For example, if we choose $c=\frac{1}{2}$ then $\gamma_2=1$ and $\gamma_1=0$ giving the method

$$y_1 = y_0 + hf(x_0 + \frac{h}{2}, y_0 + \frac{h}{2} f(x_0, y_0))$$

which is known as the *modified* Euler method. By choosing $c=1$ we force $\gamma_1 = \gamma_2 = \frac{1}{2}$ and the method becomes

$$y_1 = y_0 + \frac{h}{2} [f(x_0, y_0) + f(x_0 + h, y_0 + hf(x_0, y_0))]$$

which is known as the *improved* Euler method.

This lack of uniqueness is typical of the derivation of all Runge–Kutta methods. If we want to derive a third-order method, then we will have to solve four equations, this time in six unknowns, giving a two-parameter family of solutions. Sometimes the parameters are chosen to minimise part of the truncation error but, historically, the choice of the free parameters has usually been taken to make the coefficients of the method as simple as possible. This was done because, at the time, most of the calculations were carried out on mechanical calculators and performing these calculations was more difficult than with modern electronic calculators. For this reason the following fourth-order four-stage Runge–Kutta method was extremely popular because its coefficients are simple.

$$y_1 = y_0 + \frac{h}{6} (K_1 + 2K_2 + 2K_3 + K_4)$$

$$K_1 = f(x_0, y_0)$$

$$K_2 = f(x_0 + \frac{h}{2}, y_0 + \frac{h}{2} K_1) \qquad (8.45)$$

$$K_3 = f(x_0 + \frac{h}{2}, y_0 + \frac{h}{2} K_2)$$

$$K_4 = f(x_0 + h, y_0 + hK_3)$$

Fourth-order four-stage methods have a further advantage; they are the most accurate methods for which the number of stages is the same as the order. For an n-stage method where $n \geq 5$, the order is strictly less than n; for example, the maximum order of a 5-stage method is four. In general, the order is difficult to determine but Lambert (1973, p.122) quotes the following results which have been proved by Butcher (see references in Lambert).

Theorem 8.7

Let $p^*(n)$ be the highest order that can be attained by an n-stage method.

Then

$$p^*(n) = n \qquad n = 1,2,3,4$$

$$p^*(n) = n-1 \qquad n = 5,6,7$$

$$p^*(n) = n-2 \qquad n = 8,9$$

$$p^*(n) \leqslant n-2 \qquad n \geqslant 10$$

As an example of the use of Runge–Kutta methods we obtain the solution, at $x=0.1$, of the initial-value problem

$$y' = e^{2x} + y, \qquad y(0) = 1$$

The right-hand side of the differential equation does satisfy a Lipschitz condition and so the solution is unique; in fact, it is trivial to verify that the solution is $y=e^{2x}$. We use the fourth-order method (8.45) to obtain our approximation to the solution. Here we take $h=0.1$ so that we calculate the solution using only one step. The formulae of (8.45) give the values

$$K_1 = f(x_0, y_0) = e^0 + 1 = 2$$

$$K_2 = f(x_0 + \frac{h}{2}, y_0 + \frac{h}{2} K_1) = e^{0.1} + 1 + 0.1 = 2.20517092$$

$$K_3 = f(x_0 + \frac{h}{2}, y_0 + \frac{h}{2} K_2) = e^{0.1} + 1 + 0.1(1.10258546)$$

$$= 2.21542946$$

$$K_4 = f(x_0+h, y_0+hK_3) = e^{0.2} + 1 + 0.1(2.21542946) = 2.44294570$$

and so

$$y_1 = y_0 + \frac{h}{6}(K_1 + 2K_2 + 2K_3 + K_4) = 1.22140244$$

To eight decimal places, the correct answer is

$$y(0.1) = e^{0.2} = 1.22140276$$

8.4.2 Stability

We can say immediately that the concept of relative stability, introduced in Section 8.2.5, has no relevance here because the difference equation has only one root. We are therefore interested in the absolute stability of Runge–Kutta methods. We use the techniques of Section 8.2.5 and so we discuss stability with reference to the test equation

$$y' = \lambda y$$

The analysis is based on Lambert (1973, p.135) and we illustrate it by considering the general two-stage Runge–Kutta method (8.40). Applying this method to the test equation gives the results

$$K_1 = \lambda y_0$$

$$K_2 = \lambda(y_0 + h\alpha_2[\lambda y_0]) = \lambda y_0(1 + h\alpha_2\lambda)$$

and so

$$y_1 - y_0 = h[\gamma_1\lambda y_0 + \gamma_2\lambda y_0(1 + h\alpha_2\lambda)] = h\lambda[(\gamma_1 + \gamma_2) + \gamma_2\alpha_2 h\lambda]y_0$$

If, as in Section 8.2.5, we let $\bar{h}=h\lambda$, then we can rearrange this equation to give

$$y_1/y_0 = 1 + (\gamma_1 + \gamma_2)\bar{h} + \gamma_2\alpha_2\bar{h}^2$$

whose general solution is

$$y_n = d_1 r_1^n$$

where

$$r_1 = 1+(\gamma_1 + \gamma_2)\bar{h} + \gamma_2\alpha_2\bar{h}^2 \tag{8.46}$$

If the Runge–Kutta method is consistent then, from (8.43), $\gamma_1+\gamma_2=1$ and so (8.46) takes the form

$$r_1 = 1 + \bar{h} + O(\bar{h}^2)$$

Thus for small $\bar{h}>0$, $r_1>1$ and, as for linear multistep methods, the interval of absolute stability must have the form $(\alpha,0)$. Further, if the two-stage method is to have order two, then equations (8.43) and (8.44) must hold and (8.46) can be written as

$$r_1 = 1 + \bar{h} + \tfrac{1}{2}\bar{h}^2$$

A careful examination of this equation shows that $|r_1|<1$ if $\bar{h}$ lies in the interval $(-2,0)$. We have not specified a particular two-stage method and so this result holds for *all* two-stage Runge–Kutta methods of order two; they all have $(-2,0)$ as interval of absolute stability. In fact, for a given p ($p=1,2,3,4$), all p-stage pth-order Runge–Kutta methods have the same interval of absolute stability. When $p>4$ this is no longer the case and the free parameters of a p-stage method can be chosen to enlarge the interval of absolute stability.

8.4.3 Error control

The methods that we discussed in Chapter 7 are appropriate here. First, we consider the use of Richardson extrapolation and recall the localising assumption that no previous errors have been incurred. For a pth-order Runge–Kutta method we can write

$$y(x_1) - y_1 = \psi(x_0,y(x_0))h^{p+1} + O(h^{p+2}) \tag{8.47}$$

If we now compute $\hat{y}_1$, a second approximation to $y(x_1)$, using the same method applied at x_{-1} with a step length of $2h$, then we obtain

$$y(x_1) - \hat{y}_1 = \psi(x_{-1},y(x_{-1}))(2h)^{p+1} + O(h^{p+2})$$

We expand the right-hand side of this equation about the point $(x_0,y(x_0))$ to obtain

$$y(x_1) - \hat{y}_1 = \psi(x_0,y(x_0))(2h)^{p+1} + O(h^{p+2}) \tag{8.48}$$

Now, subtracting (8.47) from (8.48), we obtain

$$y_1 - \hat{y}_1 = (2^{p+1} - 1)\psi(x_0,y(x_0))h^{p+1} + O(h^{p+2})$$

and so the principal local truncation error, which can be used as an estimate of the local truncation error, is

$$\psi(x_0,y(x_0))h^{p+1} = \frac{y_1 - \hat{y}_1}{2^{p+1} - 1}$$

Thus, to make use of this result, we calculate values for the solution over two steps of size h; the solution is then recomputed using *one* step of size $2h$. The difference between these values, divided by $2^{p+1}-1$, is then an estimate of the local truncation error.

This estimate is sufficiently accurate to be useful for error control, but suffers from the major disadvantage that it requires a large amount of computation. This increase in effort compares badly with that necessary when using Milne's device (8.33) for predictor–corrector methods; Milne's device requires a negligible amount of computation whereas, for the Runge–Kutta method, we are considering a complete recalculation of the solution. This difference between the two approaches prompts us to ask whether some form of Milne's device can be derived for Runge–Kutta methods.

Predictor–corrector methods are a combination of *two* linear multistep methods; can we mimic the effect of Milne's device using two different Runge–Kutta methods? The answer is 'yes'. We illustrate how the error estimate can be obtained by considering the Euler and modified Euler methods. Euler's method can be written as the Runge–Kutta method

$$K_1 = f(x_0,y_0), \qquad y_1 = y_0 + hK_1 \tag{8.49}$$

When applied to the initial value problem

$$y' = e^{2x} + y, \qquad y(0) = 1$$

Euler's method gives

$$y(0.1) = 1 + 0.1\times 2 = 1.2$$

The local truncation error for this method is

$$T_1 = y(x_1) - y(x_0) - hf(x_0,y(x_0))$$

and is of $O(h^2)$. The modified Euler method

$$\hat{K}_1 = f(x_0,\hat{y}_0)$$

$$\hat{K}_2 = f(x_0+h,\hat{y}_0+h\hat{K}_1)$$

$$\hat{y}_1 = \hat{y}_0 + \frac{h}{2}(\hat{K}_1 + \hat{K}_2)$$

has local truncation error $\hat{T}_1$ of $O(h^3)$ and gives

$$\hat{y}_1 = 1.2210701$$

If we now make the assumption that

$$y_0 \simeq y(x_0) \simeq \hat{y}_0$$

then, using (8.49), we obtain the result

$$\begin{aligned} y(x_1) - y_1 &= y(x_1) - y_0 - hf(x_0,y_0) \\ &\simeq y(x_1) - y(x_0) - hf(x_0,y(x_0)) \\ &= T_1 \end{aligned}$$

Therefore

$$T_1 \simeq [y(x_1) - y_1]$$

$$= [y(x_1) - \hat{y}_1] + [\hat{y}_1 - y_1]$$

$$\simeq \hat{T}_1 + [\hat{y}_1 - y_1]$$

Now, T_1 is $O(h^2)$, whereas $\hat{T}_l$ *is* $O(h^3)$, and so the significant part of T_1 must be associated with the term $\hat{y}_1 - y_1$. Thus

$$T_1 = \hat{y}_1 - y_1 \tag{8.50}$$

and this can be used as an error estimate to control the local error of Euler's method. If the error is too large, the step length can be reduced but, if the error is very small, the step length can be increased. Johnson and Riess (1982) discuss the choice of the 'best' step length h given an error tolerance per unit step, ϵ. If we use a pth-order method and restrict ourselves to halving and doubling the step length, one strategy is to halve the step length if the estimate of the local error exceeds ϵh, double the step length if the estimate is less than $\epsilon h/2^{p+1}$ and, otherwise, retain the current step length.

For the initial value problem

$$y' = e^{2x} + y, \qquad y(0) = 1$$

we have (taking h=0.1)

$$T_1 = 1.2210701 - 1.2 = 0.0210701$$

To four decimal places, the actual error incurred is

$$1.2214 - 1.2 = 0.0214$$

and this is of the same order of magnitude as T_1.

While the generalisation of (8.50) to higher-order methods requires less computation than the use of Richardson extrapolation, it still compares unfavourably with Milne's device. For example, if we combine a four-stage fourth-order method with a six-stage fifth-order method to obtain an estimate, then we must compute the function value $f(x,y)$ *ten* times; predictor–

corrector methods, in **PECE** mode, require only *two* function evaluations. There have been many attempts to resolve this difficulty; the most successful to date has been that proposed by Fehlberg (1970). His approach, also used by others (see Lambert), is to construct a five-stage fourth-order formula using the first five K_i of a six-stage fifth-order method, thus reducing the number of function evaluations from ten to six. The resulting method, often referred to as *RKF45*, is the following:

$$K_1 = f(x_0, y_0)$$

$$K_2 = f(x_0 + \frac{h}{4}, y_0 + \frac{h}{4} K_1)$$

$$K_3 = f(x_0 + \frac{3h}{8}, y_0 + h[\tfrac{3}{32}K_1 + \tfrac{9}{32}K_2])$$

$$K_4 = f(x_0 + \frac{12h}{13}, y_0 + h[\tfrac{1932}{2197}K_1 - \tfrac{7200}{2197}K_2 + \tfrac{7296}{2197}K_3])$$

$$K_5 = f(x_0 + h, y_0 + h[\tfrac{439}{216}K_1 - 8K_2 + \tfrac{3680}{513}K_3 - \tfrac{845}{4104}K_4])$$

$$K_6 = f(x_0 + \frac{h}{2}, y_0 + h[-\tfrac{8}{27}K_1 + 2K_2 - \tfrac{3544}{2565}K_3 + \tfrac{1859}{4104}K_4 - \tfrac{11}{40}K_5])$$

The fourth-order update is

$$y_1 = y_0 + h(\tfrac{25}{216}K_1 + \tfrac{1408}{2565}K_3 + \tfrac{2197}{4104}K_4 - \tfrac{1}{5}K_5) \qquad (8.51)$$

and the fifth-order formula is

$$\hat{y}_1 = y_0 + h(\tfrac{16}{135}K_1 + \tfrac{6656}{12825}K_3 + \tfrac{28561}{56437}K_4 - \tfrac{9}{50}K_5 + \tfrac{2}{55}K_6) \qquad (8.52)$$

Subtracting (8.51) from (8.52) gives the following estimate of the local truncation error of the fourth-order method:

$$\hat{y}_1 - y_1 = h(\tfrac{1}{360}K_1 - \tfrac{128}{4275}K_3 - \tfrac{2197}{75240}K_4 + \tfrac{1}{50}K_5 + \tfrac{2}{55}K_6) \qquad (8.53)$$

This Runge–Kutta–Fehlberg method and others of similar type form the basis of several well-known ordinary differential equation solvers in program libraries. Gladwell and Sayers (1980) discuss their implementation.

We illustrate the use of (8.53) by evaluating an estimate of $y(0.1)$ for the initial-value problem

$$y' = e^{2x} + y, \qquad y(0) = 1$$

Again we take only one step and so use the method with $h=0.1$. We leave it as an exercise for the reader to verify that, to seven decimal places,

$$K_1 = 2 \qquad K_2 = 2.1012711$$

$$K_3 = 2.1557324 \qquad K_4 = 2.4058997$$

$$K_5 = 2.4437430 \qquad K_6 = 2.2100912$$

The estimate of $y(0.1)$ given by the fourth-order is

$$\begin{aligned} y_1 &= 1 + 0.1[\tfrac{25}{216}(1) + \tfrac{1408}{2565}(2.1012711) \\ &\quad + \tfrac{2197}{4104}(2.4058997) - \tfrac{1}{5}(2.4437430)] \\ &= 1.2214028 \end{aligned}$$

The estimate (8.53) for the local truncation error is

$$\begin{aligned} T_1 &= \hat{y}_1 - y_1 \\ &= 0.1[\tfrac{1}{360}(2) - \tfrac{128}{4275}(2.1557324) - \tfrac{2197}{75240}(2.4058997) \\ &\quad + \tfrac{1}{50}(2.4437430) + \tfrac{2}{55}(2.2100912)] \\ &= 0.0000000552587 \end{aligned}$$

and we can see that, according to the error estimate, our calculation gives the answer correct to about six decimal places. The exact solution is $y=e^{2x}$ and the value at 0.1 is approximately 1.2214028 which, to the given number of figures, is indistinguishable from our calculated value.

```
procedure RKF45 (function f (x, y : real) : real;
                 x0, y0, x1, epsilon, firsth, minh : real;
                 var y1, lasth : real;   var success : boolean);

  const
    assumedzero = 1E-20;

var
  k1, k2, k3, k4, k5, k6, err, x, y, h, epsilonover32 : real;
  state : (stepping, x1reached, htoosmall);
  shortrange : boolean;
```

```
begin {Runge–Kutta–Fehlberg}
  epsilonover32 := epsilon/32;   h := firsth;
  x := x0;   y := y0;   shortrange := false;
  state := stepping;
  repeat
    k1 := f(x, y);
    k2 := f(x + h/4, y + h * k1/4);
    k3 := f(x + 3 * h/8, y+h*(3*k1+9*k2)/32);
    k4 := f(x +12*h/13, y+h*(1932*k1−7200*k2+7296*k3)/2197);
    k5 := f(x+h, y+h*(439*k1/216−8*k2+3680*k3/513−
                                                845*k4/4104));
    k6 := f(x+h/2, y+h*(−8*k1/27+2*k2−3544*k3/2565
                                   +1859*k4/4104−11*k5/40));

    err := h*abs (k1/360−128*k3/4275−2197*k4/75240
                                            +k5/50+2*k6/55);

    if err > h*epsilon then
    begin
      h := h/2;
      if h < minh then state := htoosmall
    end else
    begin
      x := x + h;
      y := y + h*(25*k1/216+1408*k3/2565+2197*k4/4104−k5/5);
      if abs(x1−x) <= assumedzero then state := x1reached else
      begin
        if err <= h*epsilonover32 then h := h*2;
        shortrange := h > x1−x;
        if shortrange then
        begin
          lasth := h;   h := x1−x
        end
      end
    end
  until state <> stepping;

  if not shortrange then lasth := h;
  success := state = x1reached;
  if success then y1 := y
end {Runge–Kutta–Fehlberg};
```

Figure 8.3

The procedure *RKF45* of Fig. 8.3 implements the Runge–Kutta–Fehlberg method. An error tolerance per unit step, ϵ, must be supplied and the step length is doubled if the error estimate is less than $\epsilon h/2^5$ (because the method is fourth order) and is halved if it is greater than ϵh.

The procedure accepts an initial value for h and a minimum bound for h. Should any attempt be made to reduce h below this minimum, the process terminates and the variable parameter *success* is set *false*. The procedure returns the last value recommended for h; it is careful not to return any artifically small h used for a final short range when close to x_1. If the process is a success, the procedure also returns the computed value of y_1.

Assuming appropriate Pascal environments, the program fragments shown as Figs. 8.4 and 8.5 illustrate two alternative approaches to applying the procedure *RKF45* to n equal partitions of the interval $[x_0,x_n]$.

```
h := (xn-x0)/n;   ok := true;
for i := 1 to n do
  if ok then
  begin
    x1 := x0 + h;
    RKF45 (f, x0,y0, x1, epsilon, h, minh, y1, h, ok);
    if ok then
    begin
      writeln (x1:6:3, y1:12:8);
      x0 := x1;   y0 := y1
    end
  end
```

Figure 8.4

```
h := (xn-x0)/n;   i :=0;   status := integrating;
repeat
  i := i + 1;   x1 := x0 + h;
  RKF45 (f, x0,y0, x1, epsilon, h, minh, y1, h, ok);
  if ok then
  begin
    writeln (x1:6:3, y1:12:8);
    if i=n then status := atendofinterval else
    begin
      x0 := x1;   y0 := y1
    end
  end else
    status := failed
until status <> integrating
```

Figure 8.5

Although the Runge–Kutta–Fehlberg type of method does reduce the number of function evaluations that must be made to obtain an error estimate it is still not comparable with the convenience of the predictor–corrector estimates. However, as we have mentioned before, we can change the step length easily using Runge–Kutta methods but not at all easily using predictor–corrector methods. We discuss the relative merits of the two types of method in the next section.

8.5 Comparison of R–K and P–C methods

To make a fair comparison of Runge–Kutta and predictor–corrector methods is not simple; we first have to decide on the criteria by which they are to be judged. Topics which should be considered are accuracy, stability, cost of using the methods (measured in terms of numbers of function evaluations) and ease of programming. Lambert discusses these in some detail and we shall draw from his analysis. He points out that these properties of the methods are interrelated; for example, can we justifiably compare the accuracy of two methods when one of them uses two function evaluations and the other six? Perhaps the property that is least complicated is the ease of programming. Here the Runge–Kutta methods are seen to be better; the ease with which the step length can be changed and the lack of need for starting values count in their favour. However, we should not lose sight of the fact that the need for extra function evaluations to determine error estimates makes them less efficient than predictor–corrector methods.

Lambert concludes that if we compare only methods of the same order using the same number of function evaluations per interval, then it would appear that predictor–corrector methods are to be preferred because their stability properties are superior to those of Runge–Kutta methods. Then, if we decide to use predictor–corrector methods to determine the solution to our initial–value problem, Runge–Kutta methods are ideally suited to calculating the necessary starting values.

The choice of the method to use is not clear-cut; in special circumstances one method may be obviously better than another but the situation could be reversed if a different problem were to be considered.

8.6 Exercises

1 Write a procedure equivalent to that of Fig. 8.2 but using Euler's method (8.14) as predictor and the trapezoidal rule (8.17) as corrector. Apply your procedure to determine the solution of the initial-value problem

$$y' = 4xy^{1/2}, \qquad y(0) = 1$$

for x in the interval $[0,1]$, using step lengths $h = 0.25$ and 0.1.

2 Using Milne's device, determine an estimate of the principal local truncation error of the predictor–corrector pair of Exercise 1. Incorporate this estimate in your procedure from Exercise 1 and use your modified procedure to solve the differential equation of Exercise 1, with an error tolerance per unit step of ϵ=0.00005, at the points x=0.25, 0.5, 0.75, 1. (This predictor–corrector pair consists of two *one*-step methods and so changing the step length (halving or doubling it) presents no problems.)

3 Apply the Runge–Kutta–Fehlberg procedure of Fig. 8.3 to the initial-value problem of Exercise 1 to determine the solution at $x = 0.1, 0.2$ and 0.3.

4 Write a procedure to apply the Milne–Simpson predictor–corrector method in the mode **PECE** to estimate the solution at a set of equidistant points in the interval $[x_0, x_m]$. The procedure heading is to take the form

```
procedure MSrange (function f (x, y : real) : real;
                        x0, xm : real;   m : subs;   var y : vectors);
```

and is to assume the following environment.

```
const
   n = . . .;
type
   subs = 0 . . n;
   vectors = array [subs] of real;
```

The initial values y_0, y_1, y_2 and y_3 are assumed to be stored as the first four elements of the vector y.

Apply your procedure to the initial-value problem of Exercise 1 to produce the solution at x = 0.4, 0.5, . . ., 1.0, and use the results obtained from Exercise 3 as the initial values.

5 Compare your answers from Exercises 1, 2, 3, and 4 with the exact solution

$$y(x) = (1 + x^2)^2$$

BIBLIOGRAPHY

Atkinson, K.E. (1978), *An Introduction to Numerical Analysis*; New York: John Wiley and Sons.

Atkinson, L.V. (1980), *Pascal Programming*, London: John Wiley and Sons.

Atkinson, L.V. (1981), *Programming in Pascal* (video series), London: John Wiley and Sons.

Atkinson, L.V. (1982), *A Student's guide to Programming in Pascal*, London: John Wiley and Sons.

Birkhoff, G. and C. de Boor (1964), 'Error Bounds for Spline Approximation' *J. Math. Mech.* **13**, 827–35.

BSI (BS6192: 1982), *Specification for Computer Programming Language Pascal.* London: British Standards Institution.

Burden, R.L., J.D. Faires and A.C. Reynolds (1981), *Numerical Analysis*, Boston, Mass.: Prindle, Weber and Schmidt.

Fehlberg, E. (1970), 'Klassische Runge–Kutta Formeln vierter und niedrigerer Ordnung mit Schrittweiten–Kontrolle und ihre Anwendung auf Wärmeleitungs-probleme' *Computing* **6**, 61–71.

Flett, T.M. (1966), *Mathematical Analysis*, London: McGraw-Hill.

Gerald, C.F. (1978), *Applied Numerical Analysis*, Reading, Mass.: Addison-Wesley.

Gladwell, I. and D.K. Sayers (1980), *Computational Techniques for Ordinary Differential Equations*, London: Academic Press.

Gourlay, A.R. and G.A. Watson (1973), *Computational Methods for Matrix Eigenproblems*, London: John Wiley and Sons.

Henrici, P. (1962), *Discrete Variable Methods in Ordinary Differential Equations*, New York: John Wiley and Sons.

Isaacson, E., and H.B. Keller (1966), *Analysis of Numerical Methods*, New York: John Wiley and Sons.

Jacobs, D. (ed.) (1977), *The State of the Art in Numerical Analysis*, London: Academic Press.

Johnson, L.W. and R.D. Riess (1982), *Numerical Analsysis*, Reading, Mass.: Addison-Wesley.

Lambert, J.D. (1973), *Computational Methods in Ordinary Differential Equations*, London: John Wiley and Sons.

Leslie, P.H. (1948), 'Some Further Notes on the Use of Matrices in Population Dynamics' *Biometrika* **35**, 213–45.

McKeown, G.P. and V.J. Rayward-Smith (1982), *Mathematics for Computing,* London: Macmillan.

Ortega, J.M. (1972), *Numerical Analysis — A Second Course*, New York: Academic Press.

Ostrowski, A.M. (1973), *Solution of Equations in Euclidean and Banach Spaces*, New York: Academic Press.

Powell, M.J.D. (1981), *Aproximation Theory and Methods*, Cambridge, England: Cambridge University Press.

Ralston, A. and P. Rabinowitz (1978), *A First Course in Numerical Analysis*, New York: McGraw-Hill.

Rice, J.R. (1964), *The Approximation of Functions*, Vol.1, Reading, Mass.: Addison-Wesley.

Rudin, W. (1964), *Principles of Mathematical Analysis*, New York: McGraw-Hill.

Stroud, A.H. and D. Secrest (1966), *Gaussian Quadrature Formulas*, Englewood Cliffs, N.J.: Prentice-Hall.

Wilkinson, J.H. (1959), 'The Evaluation of Zeros of Ill-conditioned Polynomials' *Numerische Mathematik* **1**, 150–66, 167–80.

Wilkinson, J.H. (1963), *Rounding Errors in Algebraic Processes*, Englewood Cliffs, N.J.: Prentice-Hall.

Wilkinson, J.H. (1965), *The Algebraic Eigenvalue Problem*, Oxford, England: Oxford University Press.

Wilkinson, J.H. and C. Reinsch (1971), 'Handbook for Automatic Computation' Vol. II, *Linear Algebra*, New York: Springer-Verlag.

Young, D.M. (1971), *Iterative Solution of Large Linear Systems*, New York: Academic Press.

Appendix 1 OPERATORS, RESERVED WORDS AND STANDARD FUNCTIONS

Pascal Operators

Operator	*Type of left operand*	*Type of right operand*	*Type of result*	*Definition*
not		boolean	boolean	change *true* to *false* and *false* to *true*
*	real	real,int	real	multiplication
	real,int	real	real	
	integer	integer	integer	
/	real,int	real,int	real	division
div	integer	integer	integer	truncated division
mod	integer	integer	integer	remainder after **div**
and	boolean	boolean	boolean	logical 'and'
+	real	real,int	real	addition
	real,int	real	real	
	integer	integer	integer	
−	real	real,int	real	subtraction
	real,int	real	real	
	integer	integer	integer	
or	boolean	boolean	boolean	logical 'or'
=	scalar	scalar	boolean	equal to
	pointer	pointer	boolean	
<>	scalar	scalar	boolean	not equal to
	pointer	pointer	boolean	
<	scalar	scalar	boolean	less than
>	scalar	scalar	boolean	greater than
<=	scalar	scalar	boolean	less than or equal to
>=	scalar	scalar	boolean	greater than or equal to
in	ordinal	set	boolean	set membership

Note: real,int means that an operand may be either real or integer. Some of the above operators can be used with operands of other types such as, for example, sets and strings.

Pascal Reserved Words

Reserved words cannot be used as identifiers.

and	**downto**	**if**	**or**	**then**
array	**else**	**in**	**packed**	**to**
begin	**end**	**label**	**procedure**	**type**
case	**file**	**mod**	**program**	**until**
const	**for**	**nil**	**record**	**var**
div	**function**	**not**	**repeat**	**while**
do	**goto**	**of**	**set**	**with**

Pascal Standard Functions

Function	*Type of parameter*	*Type of result*	*Definition*
abs	real	real	absolute value
	integer	integer	
sqr	real	real	square
	integer	integer	
sin	real,int	real	natural sine
cos	real,int	real	natural cosine
exp	real,int	real	exponential exp(x) is e^x
ln	real,int	real	natural log
sqrt	real,int	real	square root
arctan	real,int	real	arctangent
odd	integer	boolean	tests whether integer is odd or even
trunc	real	integer	truncates a real to its integer part
round	real	integer	rounds a real to the nearest integer
ord	ordinal	integer	converts to an ordinal number
chr	integer	char	converts an ordinal number to its corresponding character
succ	ordinal	ordinal	gives the succeeding value if it exists
pred	ordinal	ordinal	gives the preceding value if it exists
eoln	text	boolean	tests for end-of-line
eof	file	boolean	tests for end-of-file

Note: In the case of *sin* and *cos* the parameter must be supplied in radians. The result delivered by *arctan* is in radians.

Appendix 2 PROGRAMS, PROCEDURES AND FUNCTIONS

Figure number	
1.1	**procedure** *FindSqRoot*
1.2	**procedure** *FindSqRoot*
1.3	**program** *RootFinder*
1.5	**function** *dotprod*
1.6	**function** *dotprod*
1.8	**procedure** *PosPrint*
1.9	**procedure** *Print*
1.10	**procedure** *AddMats*
1.11	**procedure** *MultiplyMats*
1.12	**procedure** *AddMats*
1.13	**procedure** *MultiplyMats*
1.14	**procedure** *MultiplyMats*
1.15	**procedure** *AddMats*
1.16	**procedure** *AddMats*
1.17	**procedure** *Multiplymats*
3.2	**procedure** *Bisect*
3.5	**procedure** *Secant*
3.6	**procedure** *Secant*
3.7	**procedure** *FixedPoint*
3.8	**procedure** *Aitken*
3.9	**procedure** *Aitken*
3.12	**procedure** *Newton*
3.13	**function** *poly*
3.14	**procedure** *PolyNewton*
3.15	**program** *PolyNewton*
3.16	**procedure** *Bairstow*
3.17	**procedure** *TwoSimEqns*
4.1	**procedure** *Gauss*
4.2	**procedure** *SlicedGauss*
4.3	**procedure** *SlicedGauss*
4.4	**procedure** *RecordGauss*
4.5	**procedure** *PointerGauss*
4.6	**procedure** *PermedGauss*
4.7	**procedure** *LUSolve*
4.8	**program** *PivotedGauss*
4.9	**program** *PivotedLU*
4.10	**procedure** *TriDiag*
4.11	**procedure** *Jacobi*
4.12	**procedure** *GaussSeidel*
5.1	**procedure** *Power*
5.2	**procedure** *Wielandt*
5.3	**procedure** *InversePower*
5.4	**procedure** *CyclicJacobi*
5.5	**procedure** *GivensTransform*
5.6	**procedure** *RowTransform*
5.7	**procedure** *Householder*
5.8	**procedure** *QR*
5.9	**process** *MultiplyMats*
5.10	**process** *MultiplyMats*
6.1	**function** *Lagrange*
6.4	**procedure** *CubicSplines*
6.6	**procedure** *Reduce*
6.7	**procedure** *LeastSquares*
6.9	**procedure** *Remes*
7.1	**procedure** *Romberg*
7.2	**function** *int*
7.3	**function** *int*
7.4	**procedure** *TrapAdapt*
7.5	**procedure** *TrapAdapt*
7.6	**procedure** *TrapAdapt*
8.2	**procedure** *ABM*
8.3	**procedure** *RKF45*

INDEX